Advanced Composites Engineering and Its Nano-Bridging Technology

Applied Research for Polymer Composites and Nanocomposites

Advanced Composites Engineering and Its Nano-Bridging Technology

Applied Research for Polymer
Composites and Nanocomposites

Yun-Hae Kim
Korea Maritime and Ocean University, South Korea

Ri-Ichi Murakami
Chengdu University, China

Soo-Jeong Park
Korea Maritime and Ocean University, South Korea

World Scientific

NEW JERSEY · LONDON · SINGAPORE · BEIJING · SHANGHAI · HONG KONG · TAIPEI · CHENNAI · TOKYO

Published by

World Scientific Publishing Co. Pte. Ltd.

5 Toh Tuck Link, Singapore 596224

USA office: 27 Warren Street, Suite 401-402, Hackensack, NJ 07601

UK office: 57 Shelton Street, Covent Garden, London WC2H 9HE

British Library Cataloguing-in-Publication Data
A catalogue record for this book is available from the British Library.

ADVANCED COMPOSITES ENGINEERING AND ITS NANO-BRIDGING TECHNOLOGY
Applied Research for Polymer Composites and Nanocomposites

ISBN 978-981-123-531-3 (hardcover)
ISBN 978-981-123-532-0 (ebook for institutions)
ISBN 978-981-123-533-7 (ebook for individuals)

For any available supplementary material, please visit
https://www.worldscientific.com/worldscibooks/10.1142/12238#t=suppl

Desk Editor: Nur Syarfeena Binte Mohd Fauzi

Typeset by Stallion Press
Email: enquiries@stallionpress.com

Preface

The developed original principles and approaches in advanced materials and composites define the main achievements and directions of modern natural and technical sciences, technologies, techniques and industry. Direct improvement of the materials and device characteristics are based on numerous chemical, physical and mechanical studies, modern numerical approaches, and methods of mathematical modeling and physical experiment. New scientific knowledge is based on the results of these researches, which give a possibility to understand and estimate the due technological processes and transformations of structure-sensitive properties, taking place in the fabrication of novel developed materials, composites and nanocomposites.

The success of the International Conference "Physics and Mechanics of New Materials and Their Applications" (PHENMA), held every single year from 2012 to 2019, and "Advanced Materials Development and Performance" (AMDP), held in Korea Maritime and Ocean University (KMOU) in 2014, have been a great motivation for leading this publication. The chapters have been written by several prominent scholars and scientific pioneers in this field and contain their outstanding achievements. This book is intended to contribute to the leap toward convergence technology in industries encompassing aerospace, transportation and shipbuilding based on composites engineering and its nano-bridging technology. Since the core ideology of the research is based on mutual exchange and continuity across industry–academia–research centers, this publication hopes to provide some guidance as a meaningful and intensive outcome that can encompass both academic and industrial perspectives.

Based on publications by World Scientific Publishing Company through the success of the PHENMA2018 (Chair, Yun-Hae Kim) and AMDP2014 (Chair, Yun-Hae Kim) conferences, we summarized this book on *Advanced Composites Engineering and Its Nano-Bridging Technology*. This book focused on the *in situ* research, utilization and development of advanced materials based on the following areas; materials science, polymer composites, nanomaterials, structural design, fracture mechanics, industrial applications, functionalization, physics and manufacturing process. Instead of providing a broad introduction of advanced composites and their fundamental theories with nanomaterials, the book includes practical nano-bridging techniques on nanostructures, manufacturing, analysis, applications of advanced composites using the basis of the research know-how, which had been accumulated over the years by prominent experts in these areas. The *Advanced Composites Engineering and Its Nano-Bridging Technology* book is aimed at a broad knowledge for students, engineers and specialists interested and participating in R&D of modern scientists, as their reference book.

Yun-Hae Kim,
Ri-Ichi Murakami
Soo-Jeong Park
Jodo Campus, KMOU,
Republic of Korea
September 2020

List of Contributors

Chang-Wook Park Ocean ICT & Advanced Materials Technology Research Division, Research Institute of Medium & Small Shipbuilding, Busan, Republic of Korea
pcw0591@naver.com

Do-Hoon Shin Aerostructure Development Engineering, Korean Air Aero-Space Division, Busan, Republic of Korea
dohshin@koreanair.com

Jin-Cheol Ha College of Engineering, Dali University, Yunnan, China
chnhjc@naver.com

Jing Jing Zhang Department of Nano Fusion Technology, Pusan National University, Busan, Republic of Korea
jalf1314521@gg.com

Ri-Ichi Murakami Chengdu University, Sichuan, China
Anewmoon816@gmail.com

Se-Yoon Kim Major of Materials Engineering, Department of Marine Equipment Engineering, Korea Maritime and Ocean University, Busan, Republic of Korea
seyun8269@naver.com

Soo-Jeong Park Department of Ocean Advanced Materials Convergence Engineering, Korea Maritime and Ocean University, Busan, Republic of Korea
blue9069@naver.com

Sung-Min Yoon Department of Micro-Nano Mechanical Science and Engineering, Nagoya University, Japan
zakk0902@gmail.com

Sung-Won Yoon Department of Advanced Materials, Research Institute of Medium & Small Shipbuilding, Busan, Republic of Korea
ysw8114@naver.com

Sung-Youl Bae Technology Convergence Division, Korea Institute of Ceramic Engineering & Technology, Republic of Korea
bsy@kicet.re.kr

Tae-Gyu Kim Department of Nanomechatronics Engineering, Pusan National University, Busan, Republic of Korea
tgkim@pusan.ac.kr

Tianyu Yu Major of Materials Engineering, Department of Marine Equipment Engineering, Korea Maritime and Ocean University, Republic of Korea
yutianyukmou@gmail.com

Xing Yan Tan Department of Nano Fusion Technology, Pusan National University, Busan, Republic of Korea
txy511@outlook.com

Yun-Hae Kim Department of Ocean Advanced Materials Convergence Engineering, Korea Maritime and Ocean University, Busan, Republic of Korea
yunheak@kmou.ac.kr

Acknowledgment

First, I am deeply grateful to my ex-student, Dr. Soo-Jeong Park. Dr. Soo-Jeong Park's efforts in publishing this book are very significant. She not only shared the burdens of publishing with me but also gave us many insightful suggestions to improve the completeness of the book. I want to thank her again for her willingness to respond to the writing with her doctoral course research and subsequent studies. Moreover, I would like to express gratitude to Prof. Murakami and Prof. Tae-Gyu Kim. Murakami, as a co-author of this book, readily accepted and wrote a study on the fatigue behavior of composite materials. Prof. Tae-Gyu Kim made a generous contribution by editing his studies in the nano field and having many exchanges while meeting at the international conference. Deep gratitude is also felt towards Dr. Do-Hoon Shin. He plays a pioneering role in Korean Air's research team by leading national projects in the fields of thermosetting resins and thermoplastic polymer composites. He has written on aviation composites in this publication, and as a result, he has highlighted the outstanding potential of the composites industry in this writing. I especially want to express my thanks to Prof. Jin-Cheol Ha. Through his contributions, this book is filled with more fruitful content. I am grateful to Dr. Sung-Won Yoon for writing the application fields in the shipbuilding industry while building an unrivaled area of composite materials at the Research Institute of Medium & Small Shipbuilding. Dr. Chang-Wook Park has been very successful in researching slag fibers, and he has well-written studies in the field. I am also grateful to Dr. Sung-Youl Bae for writing the research applied to automobiles, despite his busy research at the Technology Convergence Division of the Korea Institute of Ceramic Engineering & Technology. The relationship with him when Mr. Sung-Min Yoon studied abroad in France has continued to this day and it has been an delightful opportunity to work together this time. I always cheer for his research and future in Japan. Tianyu Yu and Se-Yoon Kim are grateful for this beneficial work. Finally, we are deeply grateful that the World Scientific Publishing Company has allowed us to publish through the review process so that we can edit the book like this, and we wish you tremendous success in everything you do.

Yun-Hae Kim
Jodo Campus, KMOU,
Republic of Korea
September 2020

Contents

Part 1
Introduction

Advanced Composites: From What to Why

Yun-Hae Kim

*Department of Ocean Advanced Materials Convergence Engineering,
Korea Maritime and Ocean University,
Busan, Republic of Korea
yunheak@kmou.ac.kr*

The first unsaturated polyester developed by a Swedish chemist Berzelius in 1847, the first polyester resin in 1922 and glass fibers introduced by Owens Corning in 1935 are the beginnings and first records of the history of composite materials. In 1942, the US Rubber Company developed the first glass fiber reinforced polyester composites, which finally came out to the world as structural materials. In the early days, the scarcity and high cost of composite materials led to their being limited to the fields of defense and aerospace. However, as various arrangements and types of fabrics were developed, they gradually evolved into forms capable of mass production. Today, composite materials are widely used in aircraft structural components, sporting goods, civil engineering, CNG tanks, marine vessels, offshore oil field components and automotive components. Since composite materials have been developed and commercialized, their necessity can never be overemphasized. The properties of composite materials such as high strength, high stiffness, good shear properties, low density and anisotropy are no longer limited to the macroscopic view. They enable various technological innovations as well as applications to various industries.

Generally, materials have been classified into three basic groups: metals, ceramics and polymers. Recently, composites have been widely used in various industries. Composites are made of two or more materials having different physical or chemical properties, which produce a new material with characteristics different from those of the individual components. The main constituent materials of a composite are categorized as matrix and reinforcement. The matrix binds the reinforcement together and protects it from environmental and external damage, while the reinforcement imparts strength and stiffness to the matrix. A wide variety of materials are used as reinforcement. The nano-bridging technology in polymer composites requires additional nano-related processing steps, such as a pretreatment process. Nano-based composites offer many advantages such as deformation flexibility, low weight per unit volume and design versatility, which enables the design of mechanical parts and thus expands their application to various industries.

This book was designed in this context. It focuses on the *in situ* research, utilization and development of advanced materials based on the following areas; materials science, polymer composites, nanomaterials, structural design, fracture mechanics, industrial applications, functionalization, physics and manufacturing process. Instead of providing a broad introduction of advanced composites and their fundamental theories as nanomaterials, the book includes practical nano-bridging techniques on nanostructures, manufacturing, analysis, applications of advanced composites using the basis of the research know-how it had accumulated over the years by prominent experts of these areas. With those aforementioned practical experimental considerations, this book also aims to serve various concepts for applying to industrial fields; aerospace structure, shipbuilding, marine engineering, automotives and elevator composites.

This book is divided into four parts and nine chapters. Part 1 is an "Introduction," which deals with the details of the book composition. Parts 2–4 contain the contents of research technology, as follows — "Nanomaterials and Their Experimental Approach: Nanostructures, Processing and Characterization" in Part 2, "Fracture Mechanics for Advanced Composites: Polymer Composites, Laminated Composites and Nanocomposites" in Part 3, and "*In situ*

Characterization and Applications" in Part 4. Based on these parts, the chapter contains the following contents.

Chapters 1 and 2 are titled "Performance-Based Optimal Design of Multi-layered Hybrid Composites with Halloysite Nanoparticles" and "Processing of Hierarchically Distributed Halloysite Nanotube Reinforced Composites by Electrophoretic Deposition," respectively. They commonly used Halloysite Nanotubes to analyze the structural, morphological, mechanical/physical and environmental characteristics of composites and conducted applied research. In conclusion, composite materials have high potential value in improving alternative/innovative technology to overcome the material limitations between macro polymer composites and micro nanomaterials. Chapter 3, "Plasma-Treated Carbon Black Nanofiller for Improved Dispersion and Mechanical Properties in Electrospun Complex Nanofibers," is a study that uses Carbon Black nanofiller for polymer composite. The result of this chapter evaluates and analyzes the expected effects of various surface treatments, and focuses on maximizing the utilization of materials according to the surface treatment process and research on improvements. "Characteristics of Up-Cycling Fibers Using Slag: Fiberization Process, Mechanical Properties", Chapter 4, shows the value of utilizing slag material as an industrial fiber. As a result, environmental and economic value was maximized based on the keyword of recycling, and slag as a reinforcing fiber of a polymer composite material showed high potential based on its excellent mechanical/physical properties. Chapter 5, "Fatigue Strength of Fiber Reinforced Composites Made of Carbon Fibers, Glass Fibers and Other Fibers," deals with the fatigue strength and fatigue fracture characteristics of polymer composites for various fiber materials. Chapter 6, "Corrosion and Tribological Properties of Basalt Fiber Reinforced Composite Materials," investigated the mechanical/physical/environmental properties of basalt fiber reinforced polymer composites under corrosive environments. As a result, the effects of the acids/bases on the durability of the material and the failure behavior were investigated on initial damage of materials by a corrosive environment. Chapters 7–9 focus on composite materials in the *in situ* application field, titled "Application of Composites in Aerospace Structure," "Application of Composite Materials for Shipbuilding and Marine Engineering," and "Automotive and Elevator Composite Structures," respectively. They are based on an industry-friendly research. Besides, these studies include content on core industrial technologies to improve accessibility and ease of technology for industrial needs.

Part 2
Nanomaterials and Their Experimental Approach: Nanostructures, Processing and Characterization

Chapter 1

Performance-Based Optimal Design of Multi-Layered Hybrid Composites with Halloysite Nanoparticles

Soo-Jeong Park[*,‡], Sung-Min Yoon[†,§] and Yun-Hae Kim[*,¶]

*Department of Ocean Advanced
Materials Convergence Engineering,
Korea Maritime and Ocean University,
Busan, Republic of Korea
†Department of Micro-Nano Mechanical
Science and Engineering,
Nagoya University, Nagoya, Japan
‡blue9069@naver.com
§zakk0902@gmail.com
¶yunheak@kmou.ac.kr

Halloysite nanotubes (HNTs) are an eco-friendly nanomaterial and are attracting attention as a structural polymer composite material because they have the advantage of high bonding strength with a polymer, and are emerging as an alternative material for carbon nanotubes due to their excellent reinforcing effect. Thus, the importance of HNTs is highlighted in future nanotechnology and practical applications. This chapter presents an experimental approach for the design and performance enhancement of hybrid polymer composites using HNTs. The key research focuses on the investigation of the structure and properties of HNTs to optimize their dispersion process in polymer matrices and the evaluation of material stability to improve the reliability of hybrid composite materials. Furthermore, various tests to establish their unique performance were conducted based on their optimal dispersion process and structural stability in polymer matrices.

1.1 Introduction

1.1.1 *High-Performance Polymer Nanocomposites Based on Nanofiller-Fiber Hybrid Composites*

Composites are made up of two or more materials having different physical or chemical properties, which produce a new material with characteristics different from those of the individual components. The main constituent materials of a composite are categorized as matrix and reinforcement. The matrix binds the reinforcement together and protects it from environmental and external damage, while the reinforcement imparts strength and stiffness to the matrix. A wide variety of materials are used as reinforcement. The fabrication of hybrid polymer composites requires additional processing steps, such as a pretreatment process. An example is fiber reinforced polymer matrix hybrid composites in which two or more types of continuous fiber reinforcements are cross-linked in the polymer matrix, or particulate fillers are added as reinforcement along with the fiber reinforcement. Hybrid composites offer many advantages such as deformation flexibility, low weight per unit volume and design versatility, which enables the design of mechanical parts and thus expands their application to various industries.

Representatively, Al/carbon fiber reinforced polymer hybrid composites, which are widely used for producing structural parts in the shipbuilding/marine and aerospace industries, have a high strength-to-weight ratio and excellent corrosion resistance in the vertical direction.[1] The durability of metal-composite-metal hybrid structures is being actively studied to realize specific functionalities.[2] Defects formed during the bonding of two or more different materials are considered a fatal structural defect and can significantly deteriorate the mechanical/physical properties of the composite, thus making the composites vulnerable to environmental degradation. In the case of nanomaterials, even micro-scale defects can have serious consequences for composite performance. Thus, the characteristics of hybrid composites are being investigated for specific applications in various industries.

As mentioned before, composites are advanced materials with excellent physical, mechanical and thermal properties that are different from those of the individual components. The synergetic effects of the constituent materials of the composites overcome the various limitations of the individual components, thereby making composites the leading engineering materials. Polymers, metals and ceramics form the continuous phase (matrix) of composites. Among these, polymer matrices exhibit optimum durability and heat and corrosion resistance under operation conditions. Therefore, polymer matrix-based hybrid composites with high strength and good physical and chemical stability under harsh environments were developed. Although polymers have a relatively low durability, technological advancements, such as the development of polymer-based composites by the incorporation of reinforcements, have increased their application potential in high value-added industries such as aerospace, automobile, shipbuilding and offshore industries. Moreover, nanoparticle reinforced polymers have been proposed as alternatives to overcome the vulnerabilities to heat/light/oxidation, and the fluidity and excellent formability of polymers make them a good matrix material for binding the reinforcing materials in composites.[3,4]

In nanocomposites, the uniform dispersion of nanoparticle fillers in the polymer matrix is an important factor for bonding with the fiber reinforcement. That is, to fabricate high-performance nanocomposites with isotropic properties and ensure technological universality, the nanoparticle filler and polymer should form a uniform dispersion system.[5-8] Nanoparticle fillers have infinite potential in terms of durability and performance. Nanoparticles with a size of tens to hundreds of nanometers have a high specific surface area and high aspect ratio (the ratio of larger dimension to smaller dimension). In a non-uniform dispersion system of fine particles, the total surface area of the particles is relatively large and the interparticle distance is extremely short, which significantly increases the interaction between the continuous polymer phase and the dispersed nanoparticles. This increases the surface/interface energy, which makes the fine dispersion system unstable.

Generally, nanoparticles aggregate due to van der Waals attractive forces. This significantly affects the viscosity of the polymer dispersion system. Thus, a uniform dispersion of nanoparticles is necessary to prevent their aggregation. Additionally, re-aggregation due to heat shrinkage, which is the primary aggregation occurring during stirring and the secondary aggregation occurring during the curing of the polymer, should be considered. Although nanoparticle fillers improve the mechanical, thermal and chemical properties of materials, nanoparticle aggregates larger than a certain size deteriorate formability due to the formation of an environment that promotes the adhesive aggregation of nanoparticles through hydrogen bonding, moisture adsorption and other chemical bonding mechanisms on the nanoparticle surface.

In polymer nanocomposites, nanoparticles are peeled, intercalated or dispersed in polymers such as thermoplastic and thermosetting resins. The intercalation of polymer chains in nanoparticles improves the mechanical properties and/or functionality. Therefore, to facilitate intercalation, the interaction between the polymer and nanoparticles needs to be enhanced through the surface modification of nanoparticles to an extent that does not deteriorate the intrinsic properties of the particles. Additionally, various dispersion methods, such as mechanical and ultrasonic dispersion and dispersion through dispersant addition and surface treatment, need to be considered for the uniform dispersion of the surface-modified nanoparticles in the polymer matrix.

For an ideal dispersion system, nanoparticle-polymer composites with a uniform dispersion of nanoparticles in the polymer matrix exhibit a high fracture energy due to microcrack growth inhibition and a high impact strength dependent on the type of nanoparticle filler. Currently, nanoparticles contribute to improving fracture toughness without a loss of strength, change in glass transition temperature and reduction in thermal stability. Generally, fracture toughness increases through mechanisms such as crack bridging, crack deflection, plastic deformation of the polymer around the nanoparticle, and fiber breakage, separation and pull out. Therefore, increasing the contact area between the nanoparticles and the polymer, and the uniform dispersion of nanoparticles in the polymer matrix are crucial factors for improving the fracture toughness. The aggregation of nanoparticles hinders their practical application as it decreases the active area and degrades their performance. Additionally, during their bonding with the polymer and the curing of the polymer matrix, nanoparticle aggregates create internal defects such as voids, which act as stress concentration sites and result in environmental degradation of the composite. Therefore, to expand the application scope of nanoparticle fillers, much research effort is required to reduce cohesion between the nanoparticles to minimize aggregation and ensure dispersion stability.

Figure 1.1 schematically depicts the bonding of a matrix polymer with a nanoparticle filler in a polymer nanocomposite. Figure 1.1(a) and (b) depict the ideal 1:1 mutual full bonding of each polymer and nanoparticle (complete bonding of polymer with the nanoparticles)

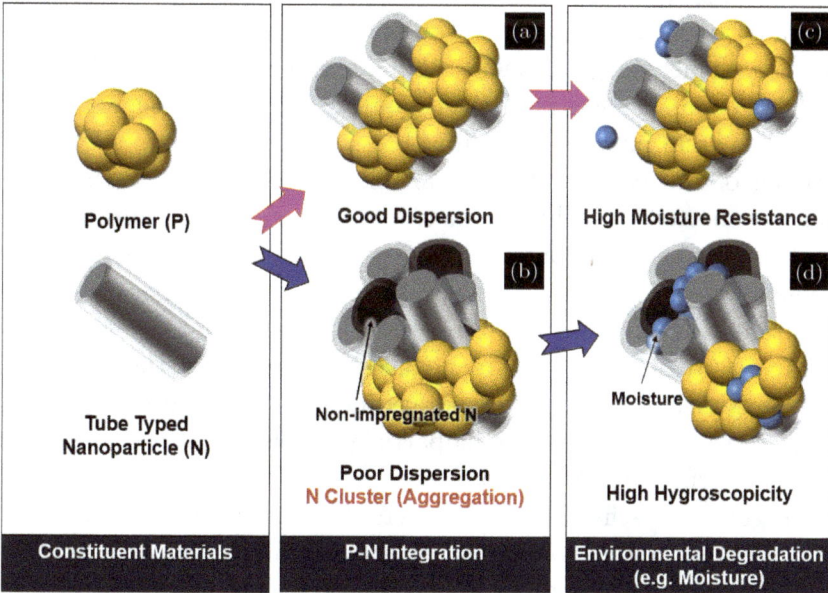

Fig. 1.1 Dispersion behavior of polymer and nanofiller. (a) Good dispersion state, (b) poor dispersion state in P-N integration, (c) high moisture resistance and (d) high hygroscopicity by P-N integration.

and the formation of aggregates in a heterogeneous dispersion, respectively. Nanoparticle aggregates do not bond strongly with the polymer and thus remain unimpregnated by the polymer. Subsequently, they are trapped as a powder material in the cured nanoparticle-polymer matrix, and their effect on the performance varies depending on the shape and size of the aggregate. Furthermore, as depicted in Fig. 1.1(c) and (d), the deterioration of polymer nanocomposites with dispersed and aggregated nanoparticles in a water environment is markedly distinct. The uniform dispersion of nanoparticles in the polymer matrix improves the stability of the polymer to environmental conditions, such as low/high temperatures, high pressures and in high humidity environments, due to the strong organic bonds between the nanoparticle and the polymer.

In contrast, nanoparticle aggregates form a non-uniform interface between the polymer and the fiber reinforcement, thereby preventing strong bonding between them, and directly participate in the swelling phenomenon and the consequent mechanical degradation. Polymer degradation by water occurs through the binding of water molecules to the polymer through chemical bonds, which exerts a tensile stress on the fiber and causes permanent damage such as fiber separation at the fiber–resin interface and fiber breakage due to the swelling and peeling of the polymer.[9] Nanoparticle aggregates, although externally connected with the polymer, internally remain unimpregnated by the polymer and contain nanoparticles in the powder form. As depicted in Fig. 1.1(c), the activation of nanoparticles by water occurs more easily than the softening of polymer by water. This is because softening of the polymer causes cracking due to the formation of a rough surface after a temporary increase in fracture toughness, whereas the aggregation of nanoparticles accelerates the damage due to short-term water activation.

Figure 1.2 schematically depicts the effect of environmental degradation on the bonding of a polymer and nanoparticles in a nanocomposite exposed to a water environment during practical application. In the case of nanoparticle aggregation, nanoparticles are activated by external environmental factors, which deteriorates the performance of the composite unlike in conventional composites.

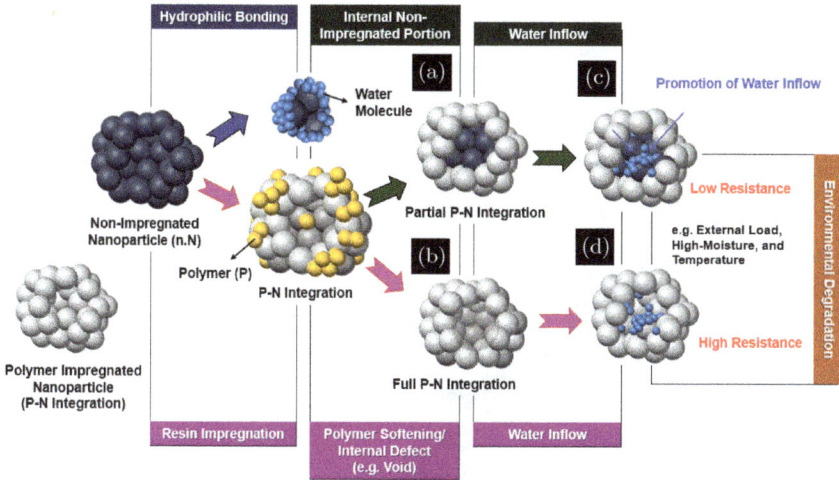

Fig. 1.2 Effect of full/partial dispersion between polymer and nanofiller on water environment degradation. (a) Partial P-N integration state (b) Full P-N integration state (c) Water inflow promotion in partial P-N integration (d) High environmental degradation resistance in Full P-N integration.

The effects of instantaneous external loads or long-term fatigue on voids and nanoparticle aggregates in a harsh water environment have a similar tendency as the stress concentration behavior for general internal defects. Additionally, nanoparticle aggregates or water-activated nanoparticle aggregates cause crack formation and growth during the composite failure mechanism, thereby showing a direct adverse effect, as depicted in Fig. 1.2(c). Thus, a hybrid technology combined with a dispersion method is required to manufacture high-performance nanocomposites.[10–12]

1.1.2 *Case Study of Halloysite Nanotubes (HNTs) for Application*[13]

The mobility (degree of freedom of movement) of these nanomaterials is also limited in the polymer and can thus be used for crack healing based on the release of healing agents through capillary action and the delivery characteristics of the nanomaterials. Figure 1.3 schematically demonstrates the change in crack profile between the

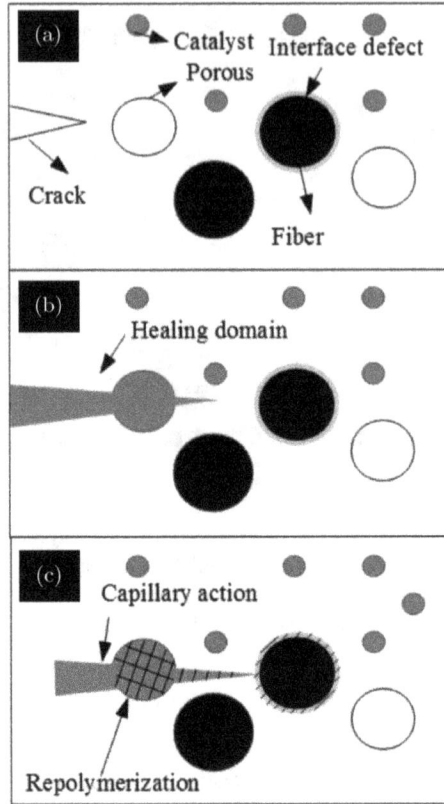

Fig. 1.3 Concept of the healing process to expect strength recovery in polymer and its interfaces; (a) initial arrangement of cracks, pores and fibers in FRP laminate, (b) healing domain and (c) repolymerization domain by capillary action.

crack interfaces as a series of steps based on the premise that the occurrence of crack healing is possible in the case of polymers under all circumstances. Generally, the breakage of composite materials includes fiber pull out, interfacial delamination and internal defects, with the formation and propagation of cracks, which eventually lead to material destruction.[14] Consequently, the incorporation of nanomaterials is also aimed at prolonging the service life and eliminating the risk of damage from external environmental factors by promoting initial crack healing.

Thus, the relationship between the recovery strength at the crack interface and the crack profile is analyzed based on factors such as the change in the area of the crack profile with crack healing.[15] The main carrier used is nanoparticles and micro-particle additives. The healing agent was released by the micro-capsules into the cracks by capillary action. To analyze the healing effect, the changes in recovery strength and crack profile with healing were predicted.[16–19]

Figure 1.4(a) depicts the model of the healing mechanism occurring through capillary action, which occurs along the precrack excepted half deformation δ in the vertical direction in Fig. 1.4(a), which is half of the deformation in the vertical direction under an

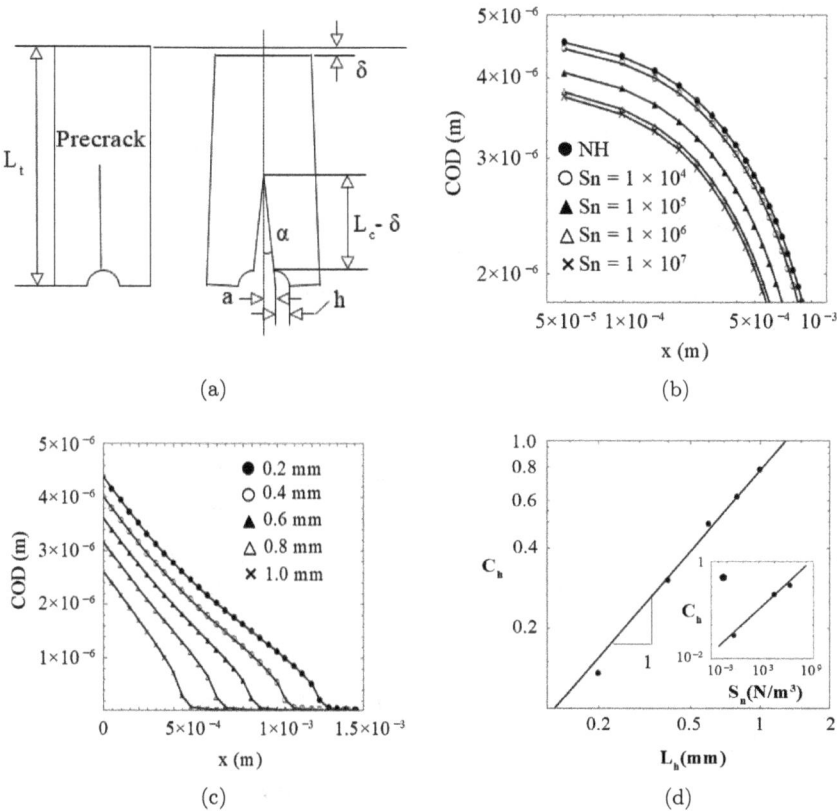

(a)

(b)

(c)

(d)

Fig. 1.4 Results of a numerical simulation on the nanocomposite to predict the healing efficiency. (a) Geometry of a cracked specimen, (b) The stiffness of interface element changes, (c) The length of healed crack changes, (d) Healing efficiency.

applied external load (Equation (1.1)).

$$\text{COD} = \frac{2}{L_c - \delta} \left\{ \frac{\delta}{L_c}(L_t - \delta) - h \right\} (L_t - \delta - x) \qquad (1.1)$$

where (a is in the Fig. 1.4(a)) is the initial half crack open displacement profile, h is the radius of the hole, L_t is the total longitudinal half length and L_c is the crack length. The angle of crack opening displacement (COD), α, is assumed to be $2\alpha = \text{COD}$. As depicted in Fig. 1.4(b), based on the fixed parameters, that is, a precrack length of 1.5 mm and a healed crack length of 0.3 mm, the interface elements, stiffness ($S_n = E_n/t_n$, where E_n is the modulus of the healed elements and t_n is the normal thickness), and size are controlled. At present, the position of the profile, which is X, is simulated by the finite element method (FEM). Figure 1.4(c) shows the change in the length of the healed crack at a fixed stiffness of 1×10^7 and a precrack length of 1.5 mm. This was used to calculate the healing efficiency ($C_h = 1 - V_h/V_{nh}$, where V_n and V_{nh} are the volumes of the healed and non-healed segments, respectively), as depicted in Fig. 1.4(d). Thus, by using a linear approximation to the aforementioned geometry depending on the constituent materials and structure of a nanocomposite, the healing efficiency can be predicted. In the case of nanocomposites, the nanostructure and nanoscale dimensions are difficult to determine reliably based on experiments. Moreover, analyzing the properties of each particle or the complex (hybridized) microstructures is virtually difficult. Therefore, with the widespread application of nanocomposites and hybrid composites, FEM modeling is considered an alternative preliminary step to experimental investigation.

1.2 HNTs Reinforced Polymer and Its Unique Features

1.2.1 Structural Characteristics of HNTs

HNTs are an environmentally friendly nanomaterial formed by the geological weathering or hydrothermal modification of rocks, saprolite and soil. The water content of HNTs is higher than that of

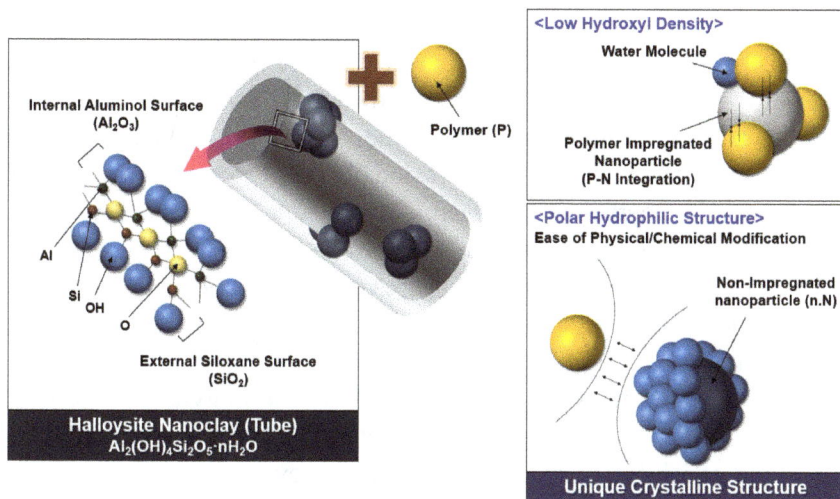

Fig. 1.5 Schematic diagram of the physical/chemical characteristics of the HNT structure.

the kaolin group minerals.[20–22] Although chemically similar to kaolinite, HNTs having the chemical formula of $Al_2(OH)_4Si_2O_5.nH_2O$ are separated by a single layer of water molecules. HNTs are hollow with an outer diameter of approximately 30–190 nm, a cylindrical pore (lumen) of approximately 10–15 nm, and an inner diameter of approximately 10–100 nm.[23, 24] As depicted in Fig. 1.5, the 1:1-type layer structure is largely divided into silica tetrahedral sheets and alumina octahedral sheets having siloxane (Si–O–Si), silanol (Si–OH) and aluminum hydroxide (Al–OH) groups, with a thin layer of water between the continuous layers.[20, 25–27] Different reactions occur on the outer surface, inner lumen surface and interlayer surface of the tube. The zeta potential of HNTs is mostly negative at pH 6–7 because of the small surface potential of SiO_2 and the presence of cations on the inner surface of Al_2O_3. The chemical properties of the outermost surface and inner cylindrical core of HNTs are similar to those of SiO_2 and Al_2O_3, respectively. HNTs have a polar hydrophilic structure because of the presence of OH groups at the end of the sheet, which makes the intercalation of a lipophilic material difficult. However, naturally occurring HNTs can be easily and uniformly dispersed in a polymer because of its unique crystalline structure with a lower hydroxyl density as compared to other nano-sized inorganic fillers. Its incorporation into a polymer matrix improves the mechanical properties of the nanocomposites, such as Young's modulus and the yield stress. Additionally, HNTs delay heat transfer and lower the coefficient of thermal expansion (CTE) by forming a physical barrier during the combustion reaction.

HNTs have a variety of structures and properties depending on the interlayer spacing.[25, 28, 29] The hydrated form of HNTs ($n = 2$) has an interlayer spacing (crystallite size) of 10 Å with the monolayers separated by water molecules, whereas the dehydrated form of HNTs ($n = 0$) has an interlayer spacing of 7 Å, which is obtained by removing the interlayer water molecules; nevertheless, the tubular morphology is maintained. Naturally occurring HNTs exist as a mixture of 10 Å and 7 Å interlayer spacing and are referred to as Natural-HNTs (Hydro-HNT) in this chapter. A Natural-HNT is transformed into meta-kaolin by dehydroxylation at an elevated temperature of 450–700°C at which the 10 Å crystal grains are removed and only the 7 Å crystal grains remain. Frequently, Natural-HNT, a mixture of 10 Å and 7 Å, and crystalline HNT with 7 Å grains are collectively referred to as endelite. Crystalline HNT transforms into amorphous HNT, which is hereafter referred to as Amorphous-HNT (Meta-HNT), at a temperature of 700°C or higher. These structural changes are caused by the dehydroxylation of water, which is associated with an important thermal dissociation reaction occurring in natural and synthetic silicate minerals of the kaolin group. The loss of the hydroxyl group causes local buckling and deformation in the crystal layer structure of HNT. Thus, distinguishing the structures of HNT based on the changes in the structure and properties of HNTs, such as particle size, morphology and defect density due to dehydroxylation reactions, remains controversial, as they depend on many factors such as heating rate, sample processing, temperature and humidity. However, these factors significantly affect the performance of HNTs, and much research is required to obtain HNTs with uniform properties and an optimal structure through the modulation of these parameters.

1.2.2   *FTIR Spectroscopy*[30]

Fourier-transform infrared (FTIR) spectroscopy was performed to clearly distinguish the structural diversity of HNTs with Fig. 1.6 depicting the results. FTIR spectroscopic analysis allows the quantitative and qualitative evaluation of substances through the infrared irradiation of a compound. When a compound is irradiated with infrared rays, the molecules absorb certain wavelengths of radiation in the energy region of 2.5–25 μm (= 4000–400 cm^{-1}) depending on the bonding between the atoms in the molecule, and the change in emission is measured. This allows for the identification of the functional groups (–OH, C=O, COOH, N–H, C=C, etc.) in the molecules forming the compound.

Figure 1.6 depicts the FTIR spectra of Natural-HNT, Crystalline-HNT and Amorphous-HNT. Generally, peaks in the wavelength range of 3700–3600 cm^{-1} correspond to the stretching vibrations of the hydroxyl groups. The peaks at 3691.84 and 3621.11 cm^{-1} in the Natural-HNT FTIR spectrum correspond to the stretching vibrations of the inner hydroxyl stretching, and the stretching

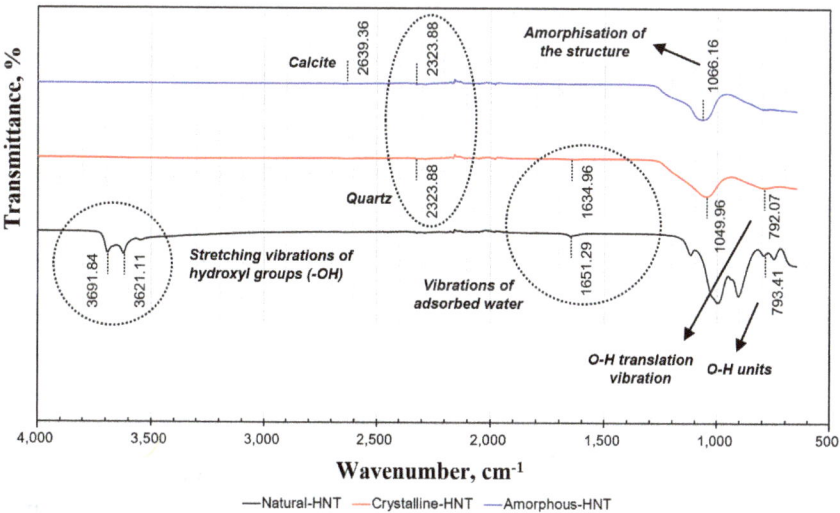

Fig. 1.6 FTIR analysis of structural characteristics according to the crystallinity of HNT.

vibration band groups, respectively. Two peaks corresponding to the Al_2OH stretching bands were observed in the FTIR spectrum of Natural-HNT but were absent in the FTIR spectra of Crystalline-HNT and Amorphous-HNT. The disappearance of these bands is because of dehydroxylation due to heat treatment. Besides, the peak at 2323.88 cm^{-1} of a newly formed Crystalline-HNT and an Amorphous-HNT indicates a quartz structure. Additionally, a peak appeared in the FTIR spectrum of Amorphous-HNT at 2639.36 cm^{-1}, which can be ascribed to a calcite structure formed by the complete transformation into an amorphous structure. The peaks at 1651.29 cm^{-1} for Natural-HNT and 1634.96 cm^{-1} for Crystalline-HNT correspond to the vibrations of adsorbed water molecules, with Crystalline-HNT showing a significantly weaker peak than that of Natural-HNT. The 1049.96 cm^{-1} peak in the FTIR spectrum of Crystalline-HNT red-shifted to 1066.16 cm^{-1} in the FTIR spectrum of Amorphous-HNT due to the formation of an amorphous structure. Furthermore, the peaks in the wavelength range of 1000–400 cm^{-1} correspond to the vibrations of the Si–O–Si, Al–O–H and OH groups. The translation vibration of the hydroxyl groups of HNTs gave rise to a peak at 793.41 and 792.07 cm^{-1} in the FTIR spectra of Natural-HNT and Crystalline-HNT, respectively. Although it has not been confirmed in the present results, the presence of more than three bands in the region of 3900–3600 cm^{-1} in the FTIR spectrum of Amorphous-HNT, which correspond to the hydrogen bonds between the various hydroxyl groups of γ-alumina, is considered a structural characteristic of amorphous HNTs.

1.2.3   XRD Analysis[30–32]

The structural deformation and changes in the crystallinity of Natural-HNT with the heat treatment temperature (heating temperature) were analyzed by X-ray diffraction (XRD) with the results depicted in Fig. 1.7. The peak at 26.6°C appeared in the XRD spectra regardless of the change in heat treatment temperature, which can be attributed to quartz. Natural-HNT contains both hydrated (10 Å interlayer spacing) and dehydrated (7 Å interlayer

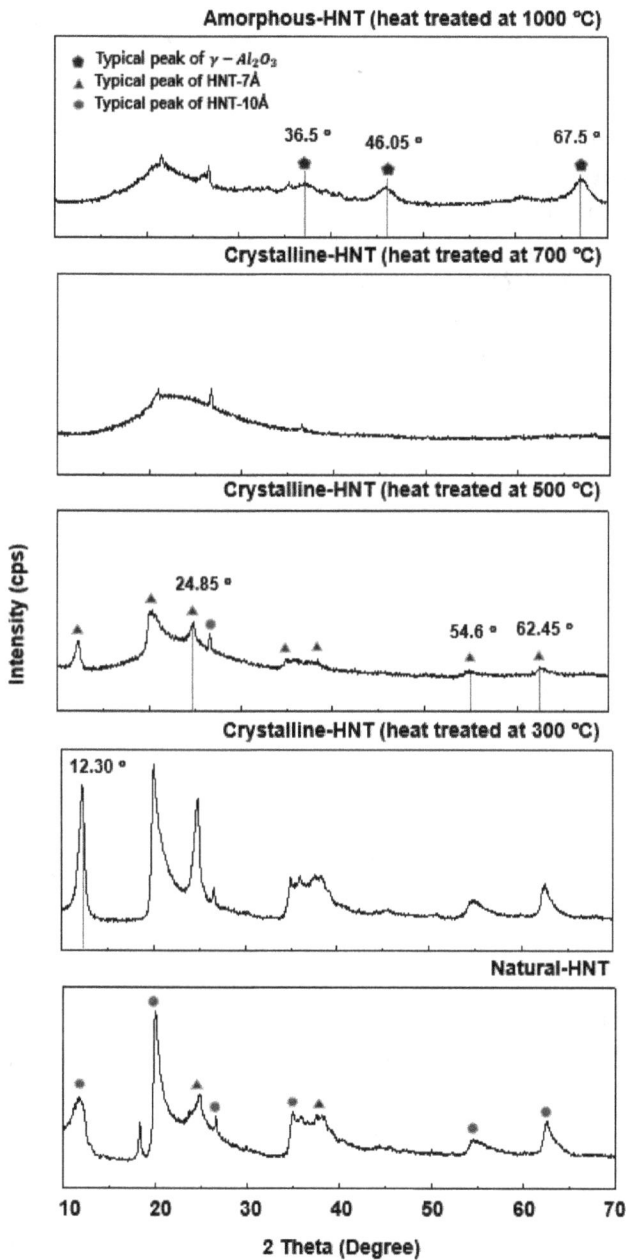

Fig. 1.7 XRD analysis of the characteristics of the HNT structure according to the changes in the heat-treatment temperature.[30–32]

spacing) forms, which changed significantly when heated above 300°C. Structural deformation occurred prominently for both the 10 Å and 7 Å grains, and above 700°C, even the 7 Å grains were destroyed and a mixture of amorphous substances with quartz or silicon oxide was formed. That is, when heated at 500–700°C, dihydroxylation occurred, which was originally due to the reduced coordination of octahedral aluminum. Additionally, an intense peak appeared at 1000°C due to the formation of an alumina-rich phase. The XRD peaks at 12.30°C and 24.85°C are the characteristic diffraction peaks of HNT. No changes were observed in these peaks until approximately 300°C; however, above 500°C, the crystalline silicate layers transformed into an amorphous structure through the dehydration and oxidation of the alumina groups. For Amorphous-HNT, the presence of mullite was considered a characteristic feature, with the possibility of cristobalite formation. The peak at 46.05°C indicates the formation of aluminum oxide, while the peak at 26.15°C indicates the formation of mullite or SiO_2.

1.2.4 TEM Micrograph[31, 32]

Figure 1.8 depicts the results of the transmission electron microscopy (TEM) analysis of the morphological changes of HNT with heat treatment. Rod-like structures were the main morphological feature in all the HNTs, and most of them existed as aggregates (clusters). Natural-HNT (untreated form) exhibited a smooth surface, whereas Crystalline-HNT and Amorphous-HNT exhibited rough surfaces with partial irregularities because of structural deformation due to heat treatment.

Specifically, Amorphous-HNT exhibited a large amount of irregularities several nanometers in size. Above 300°C, structural destruction occurred due to dehydration, resulting in the removal of adsorbed and interlayer water, and at 500°C, the silicate layer was significantly destroyed because of the removal of the OH groups due to dehydration. At 1000°C, meta-kaolin decomposed (sintering) and an amorphous structure was formed.

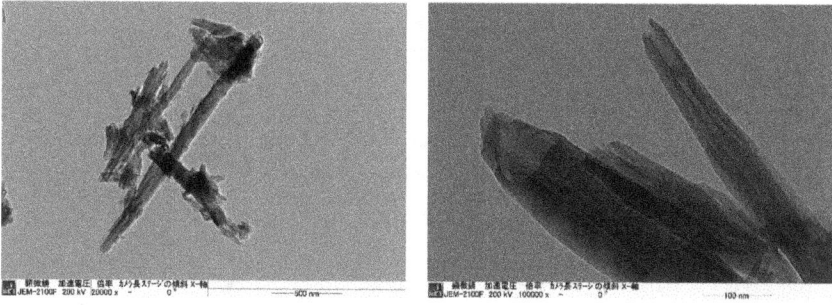

(a) Natural-HNT

(b) Crystalline-HNT (heat treated at 300°C)

(c) Crystalline-HNT (heat treated at 500°C)

Fig. 1.8 TEM observations of the morphological analysis of the HNT structure according to heat-treatment temperature at 20,000 magnification (left) and 100,000 magnification (right).

(d) Crystalline-HNT (heat treated at 700°C)

(e) Amorphous-HNT (heat treated at 1000°C)

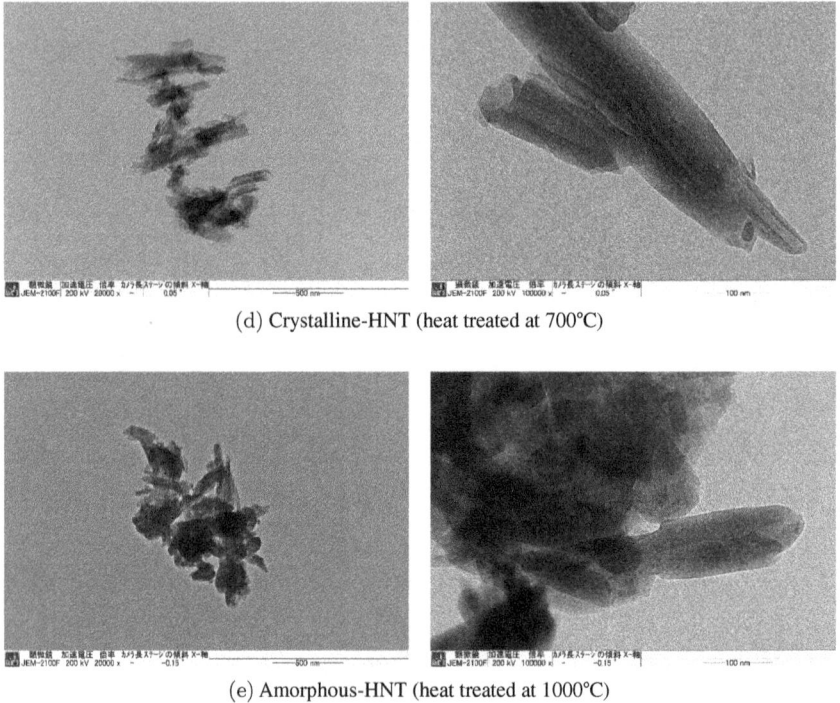

Fig. 1.8 (*Continued*)

1.3 Effect of HNTs on Thermal, Hygroscopicity and Mechanical Properties of Nanocomposites

1.3.1 *Constituent Materials of HNT Reinforced Nanocomposites and Their Properties*[30]

The excellent physical and chemical properties of HNTs, which make it an excellent affordable alternative to carbon nanotubes as a filler for polymer nanocomposites, are now discussed. Delamination and interlayer separation at the interlayer interface are the common damages that limit the service life of laminated composites. These are mainly caused by static or fatigue loading at the manufacturing stage, or stress concentration. To improve interfacial bonding, a small amount of materials, such as SiC and TiO_2, is added for reinforcement. HNTs are similarly used as a reinforcement. HNTs act as a bridge between the fiber and the polymer matrix and mainly improve the physical bonding and mechanical strength, such as the bending strength and the inter-laminar shear strength (ILSS).

Table 1.1 lists the main properties of the constituent materials of laminated nanocomposites. Two types of reinforcements, unidirectional glass and basalt fibers, a thermosetting epoxy resin matrix and HNT nanofillers were used. Here, the effect of Natural-HNT, Crystalline-HNT and Amorphous-HNT on the properties of laminated nanocomposites was intensively studied. Glass fibers and basalt fibers have similar chemical compositions, and their mechanical properties are lower than those of high-performance carbon fibers. Glass fibers are widely used in various industries, while basalt fibers are an environmentally friendly material made from basalt ore and possess excellent heat resistance, and mechanical and chemical properties. The availability of these materials is highly appreciated in industries. A low-viscosity bisphenol-A (BPA)-type thermosetting epoxy resin was used as the matrix. BPA-type epoxy resins are widely used as adhesives, coating agents and matrices of composite materials owing to their excellent chemical resistance, solvent resistance, high mechanical strength and dimensional stability. Epoxy matrix composites with fiber reinforcements are light-weight and durable, and exhibit adhesiveness and high mechanical properties, which make them extremely versatile for aerospace, automobile, civil structure, and shipbuilding and marine applications.

1.3.2   *Structural Characterization*[30]

Before manufacturing the laminated nanocomposites, the cross-linked structure was analyzed by FTIR spectroscopy to verify the uniform dispersion of HNTs in the epoxy matrix and the structural changes between epoxy and HNT with the results depicted in Fig. 1.9. The FTIR spectrum of pure epoxy (Neat EP) shows a weak but wide peak at 3502.44 cm^{-1} corresponding to the O–H stretching vibrations and a peak at 2925.98 cm^{-1} corresponding to the C–H

Table 1.1 Main specifications of constituent materials in laminated nanocomposites

Properties	Reinforcements[*]		Matrix[**]	
	Glass fiber	Basalt fiber	Properties	Epoxy
Type Plain woven fabric weight (g/m^2)	Unidirectional 450	Unidirectional 300 ± 24	**Density (g/ml)** Viscosity (cps)	1.0–1.2 (1.160) 800–1,100 (935)
Plain woven fabric thickness (mm)	0.177	0.115 ± 0.013	Glass transition temperature (T_g, °C)	70–80
Chemical Composition	E-glass SiO_2, Al_2O_3, CaO, MgO, Na_2O, K_2O, B_2O_3	SiO_2, TiO_2, Al_2O_3, Fe_2O_3 + FeO, CaO, MgO, MnO, Na_2O + K_2O, SO_3	Post curing	80 °C for 4 h

Properties	Nanofiller[***]		
Halloysite Nanoclay	Synonym Kaolin clay	Density (g/cm^3) 2.53	Molecular weight (g/mol) 294.19

Notes: [*]Glass fabric, EJ30, supplied by Hankuk Carbon Co., LTD, Republic of Korea; Basalt fabric, HB-300, supplied by GM Composite Co., Republic of Korea. [**]Epoxy resin, KFR-120V supplied by Kukdo Chemical Co., Ltd, Republic of Korea (Hardener: KFH-141). [***]Halloysite Nanoclay supplied by Sigma-Aldrich, Republic of Korea (CAS No. 1332-58-7).

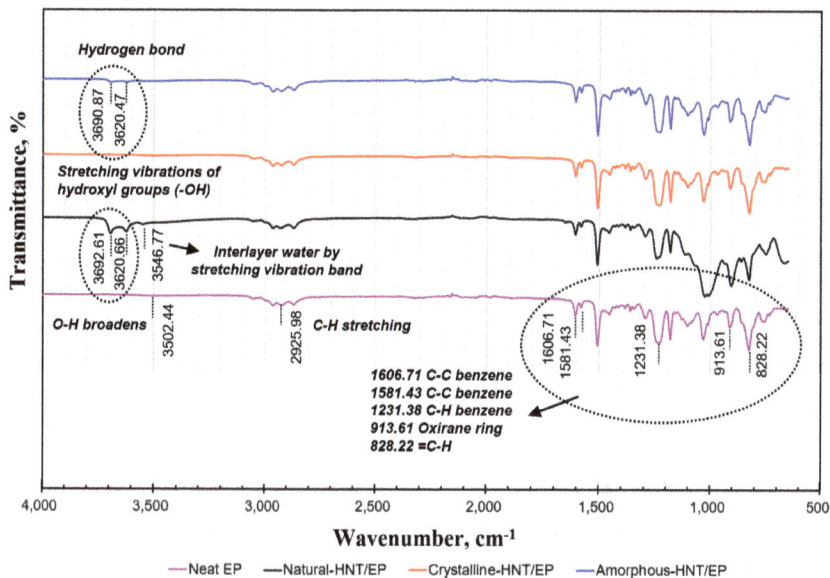

Fig. 1.9 FTIR analysis of the effect of HNT crystallinity on the chemical structure of an HNT/Epoxy colloidal solution.

stretching vibrations. As depicted in Fig. 1.9, and as compared with Fig. 1.10, the peaks at 1606.71 and 1581.43 cm^{-1} correspond to the C–C bond and the peak at 1231.38 cm^{-1} corresponds to the C–H bond in the benzene ring of epoxy, respectively. The oxirane ring at the end of the epoxy structure gave rise to a peak at 913.61 cm^{-1}, while the =C–H structure gave rise to a peak at 828.22 cm^{-1}. The addition of Natural-HNT, Crystalline-HNT and Amorphous-HNT to Neat EP resulted in distinct changes to the intermolecular bonds in the structure.

The addition of Natural-HNT gave rise to peaks at 3692.61 and 3620.66 cm^{-1} corresponding to the stretching vibrations of the hydroxyl groups. Additionally, the peak at 3546.77 cm^{-1}, corresponding to the stretching vibration of the hydroxyl group, indicates the presence of interlayer water. In contrast, the addition of Crystalline-HNT did not give rise to peaks corresponding to the stretching vibrations of the hydroxyl groups, which indicates the dehydrate structure of Natural-HNT after the high-temperature heat treatment. With the

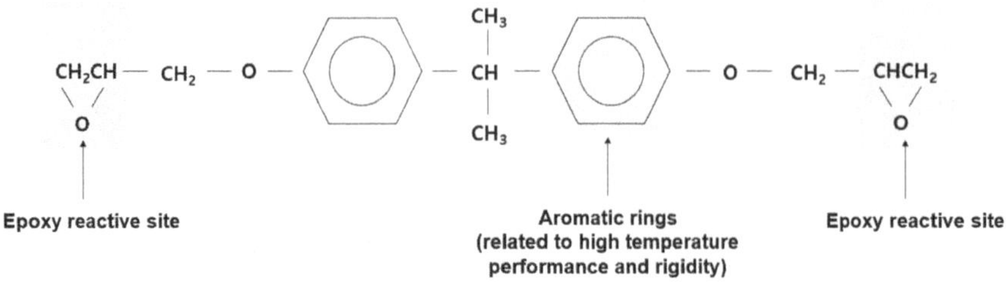

Fig. 1.10   Molecular structure of epoxy.

addition of Amorphous-HNT, peaks corresponding to the stretching vibrations of the hydroxyl groups reappeared at 3690.87 and 3620.47 cm^{-1}, which indicates the presence of hydrogen bonds.[20, 33, 34]

In conclusion, the chemical similarity of epoxy and HNTs leads to their strong bonding. Despite the strong bonding between epoxy and HNTs, some HNT aggregates are formed in the composite due to the strong cohesion between HNTs. Therefore, to utilize HNT, it is necessary to ensure the uniform dispersion of HNT in the epoxy and its dispersion stability, which can be realized through surface modification.

1.3.3   *Thermal Analysis of the Curing Characteristics*[30]

Table 1.2 lists the results of the differential scanning calorimetry (DSC) analysis on the effects of HNT crystallinity and content on the curing of the epoxy system. With an increase in the heating rate from 10°C/min to 20°C/min, the onset temperature of the endothermic/exothermic reaction increased by 38% for all the HNT types under the same conditions. Additionally, the heat of reaction increased with an increasing heating rate. Furthermore, unlike Crystalline-HNT, Amorphous-HNT did not show a certain tendency. This is because of its relatively weak structural properties, that is, its bonding with epoxy is less affected by heat. Moreover, the crystallinity of HNT affects the cohesion between the particles, and thus, the dispersion stability. Therefore, the non-uniform dispersion of HNT in epoxy delays the curing reaction, and the longer the

Table 1.2 Parameters for analyzing the thermal properties of nanocomposites according to the crystallinity and content of HNT

Samples	Filler loading (wt%)	Heating rate (°C/min)	$\Delta H_{reaction}$ (J/g)	T_{onset} (°C)	T_P (°C)	T_{endset} (°C)	Fully cured temperature (°C)
Neat Epoxy		10	69.95	52.48	122.27	181.17	180.8
		20	233.35	81.54	132.71	209.81	209.5
Crystalline-HNT/Epoxy (heat-treated at 700°C)							
0.5-Crystalline-HNT/epoxy	0.5	10	69.85	39.38	121.68	195.98	151.2
		20	168.30	92.99	138.37	190.22	189.9
1-Crystalline-HNT/epoxy	1	10	69.85	39.38	121.68	195.98	193.7
		20	142.03	92.90	136.85	192.77	192.4
3-Crystalline-HNT/epoxy	3	10	510.56	37.62	118.71	205.46	205
		20	798.41	85.25	138.86	204.47	204.5
Amorphous-HNT/Epoxy (heat-treated at 1,000°C)							
0.5-Amorphous-HNT/epoxy	0.5	10	121.98	36.94	115.92	188.82	188.5
		20	125.26	96.49	137.53	185.05	184.7
1-Amorphous-HNT/epoxy	1	10	497.78	55.01	112.93	203.57	203.2
		20	122.15	95.80	139.24	186.45	186.1
3-Amorphous-HNT/epoxy	3	10	249.84	79.99	114.55	162	161.8
		20	155.59	93.19	137.42	184.74	184.4

exposure to the activation curing temperature, the greater the heat of reaction. That is, Amorphous-HNT, which has strong van der Waals forces between the particles, easily forms clusters during its dispersion in epoxy, thus slowing the curing reaction. However, generally promoting of the curing reaction at a relatively low temperature by the addition of a nanofiller, such as HNT, leads to an improvement in thermal stability.

Figure 1.11 depicts the changes in the degree of cure with time (Equations (1.2) and (1.3)) and the rate of cure at each temperature with heating rate.

$$\% \text{ cure} = \frac{dH\,(\text{uncure}) - dH(\text{cured})}{dH(\text{uncure})} \times 100 \tag{1.2}$$

$$\alpha = \frac{\Delta H_t}{\Delta H} \tag{1.3}$$

where α is the conversion, ΔH_t is the cumulative calorific value ranging from 0 to t, and ΔH is the total calorific value when $\alpha = 1$.

The degree of cure was calculated by integrating the calorific value over time. Both the degree of cure and rate of cure increased with increasing temperature. In particular, the degree of cure shows an S-shaped curve, which indicates an autocatalytic reaction of a typical amine-based curing agent in which the curing reaction proceeds slowly after reaching the maximum rate of cure. The effects of Natural-HNT, Crystalline-HNT and Amorphous-HNT on the curing of epoxy were the same, that is, the faster the heating rate, the higher the onset curing temperature and the faster the rate of cure. The main difference was that, at a low concentration of 0.5 wt%, Crystalline-HNT decreased the curing temperature and started the curing reaction instantaneously, whereas Amorphous-HNT did not participate in the curing reaction of the epoxy, as it had a negligible effect on chemical bonding at such a low amount and was difficult to disperse uniformly due to its instability. At a concentration of 3 wt% or more, HNT, regardless of its crystallinity, inhibited curing under the same temperature conditions, which is evidenced by the short reaction time and lower onset temperature of the curing reaction.

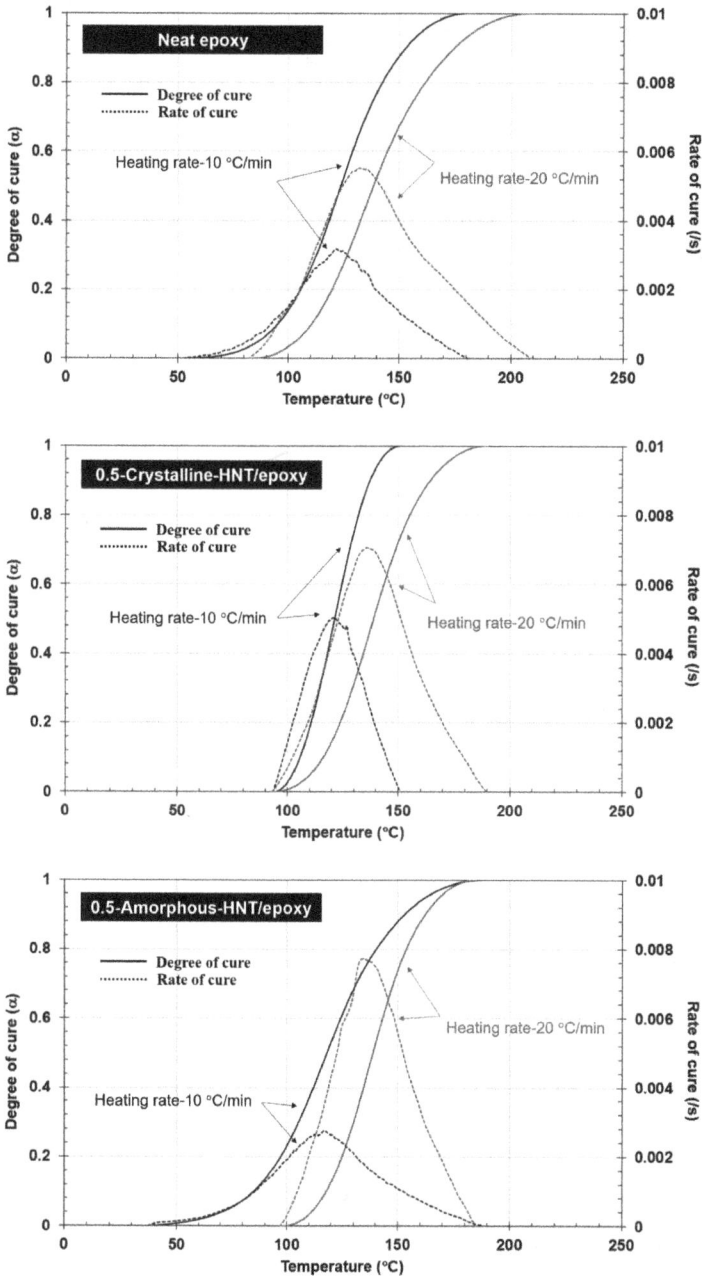

Fig. 1.11 DSC analysis of the thermal properties of nanocomposites according to the crystallinity and content of HNT with degree of cure and rate of cure.

S.-J. Park et al.

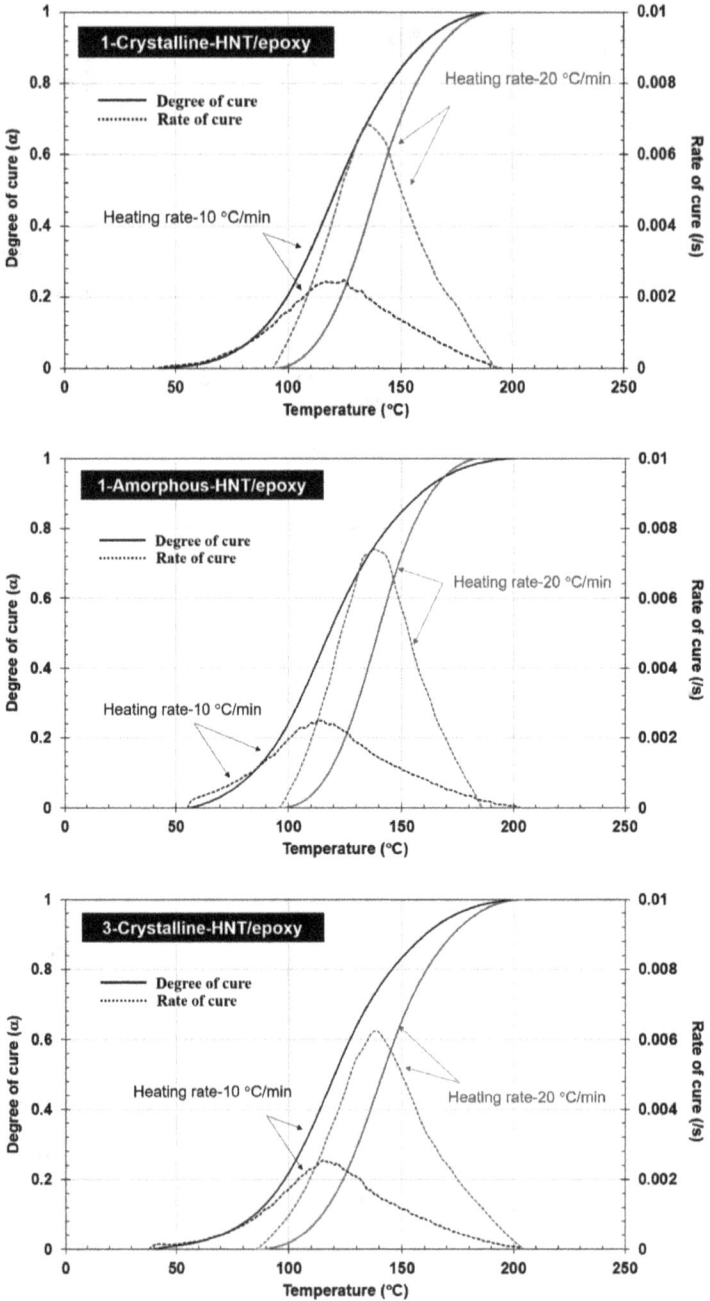

Fig. 1.11 (*Continued*)

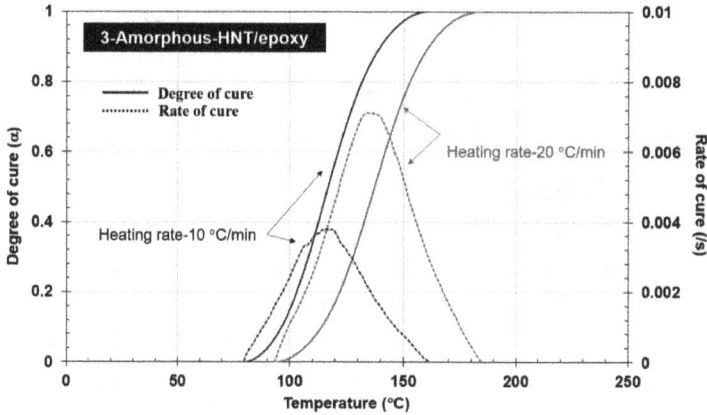

Fig. 1.11 (*Continued*)

In conclusion, regardless of its crystallinity, HNTs lowered the curing temperature until complete curing or increased the rate of cure within a range similar to that of epoxy. The incorporation of nano silica-based particles into epoxy promotes its curing due to the increased interface interaction between the nanoparticles and the matrix. Thus, in all the cases, the rate of cure decreased sharply with a decrease in the heating rate, because sufficient time was required for curing. Additionally, an increase in the heating rate allowed the curing reaction to proceed at a high temperature until curing was complete. In particular, an HNT content of 1 wt% had an insignificant effect on the curing reaction regardless of crystallinity and heating rate. At a heating rate of 10°C/min, the curing reaction proceeded for a long time and required a high temperature. Moreover, Crystalline-HNT at a concentration of 1 wt% or higher maintained almost the same curing reaction conditions as those of neat epoxy. Amorphous-HNT at a concentration of 3 wt% showed a stable curing reaction according to the heating rate and provided optimum curing conditions in terms of temperature, rate of cure and curing reaction time until complete curing. However, since the cohesion between particles is higher in the case of Amorphous-HNT than in the case of Crystalline-HNT, further research is required to achieve a stable dispersion of Amorphous-HNT.

1.3.4   Effect of HNT Crystallinity on Water Absorption and the Mechanical Properties of Glass Fiber Reinforced and Basalt Fiber Reinforced Polymer (BFRP) Nanocomposites[30]

Polymer matrix-based composites are vulnerable to environmental degradation due to moisture, solvent, oil, temperature, mechanical load and radiation. In particular, prolonged exposure to high humidity or a water environment promotes polymer degradation, which deteriorates the service life of the product. The absorbed water molecules act as a plasticizer inside the polymer and at the interface between the fiber reinforcement and the polymer matrix, thereby weakening the interfacial strength and causing micro-mechanical damage. Although the plasticization of polymer increases the fracture toughness of the nanocomposite, a prolonged exposure to moisture creates a rough surface and cracks, and causes micro-mechanical damage at the interface between the fiber reinforcement and the polymer matrix.

Initially, moisture penetrates the polymer and diffuses according to Fick's law, which is often considered the mechanism of deterioration of the polymer by moisture diffusion. The moisture diffusing into the polymer mainly penetrates the non-uniform interface and causes swelling, thereby deteriorating the properties. A large amount of moisture causes a stress in the tensile direction inside the fiber or causes permanent damage, such as interfacial separation due to swelling and peeling of the polymer, and fiber failure. Unlike in the case of a polymer matrix, direct water penetration has a relatively small effect on fiber reinforcement or surface damage, but severely damages the fiber surface through polymer deterioration. Although most of the absorbed moisture is chemically bonded to the polymer, some diffuses into the free volume and does not chemically bind to the polymer, and is removed upon drying without affecting the mechanical properties of the polymer. The swelling and drying of the polymer lead to material shrinkage, which weakens the interfacial bonding. That is, the ingress of a large amount of water or prolonged exposure to moisture causes irreversible swelling of the polymer matrix, which leads to interfacial damage; thus, the physical properties of the polymer cannot be recovered upon drying. Clay nanofillers can prevent the irreversible swelling of the polymer matrix by reducing the diffusion of water owing to their excellent barrier properties. This is an advantage since moisture also inhibits the polymerization of the polymer matrix, thereby preventing the synthesis of polymer nanocomposites. Nanofillers with a high aspect ratio, such as HNTs, improve moisture absorption by forming a tortuous path for the diffusion of water molecules into the polymer. The decrease in the cross-link density of the polymer due to the tortuosity effect significantly decreases the equilibrium moisture content and diffusion coefficient of the composites. A large amount of hydrophilic HNT (Natural-HNT) reduces the water uptake by the polymer owing to its superior moisture resistance, thereby improving the water resistance of the nanocomposite.[35]

Table 1.3 lists the results of the water absorption test (ASTM D5229) of glass fiber reinforced polymer (GFRP) and BFRP nanocomposites reinforced with Natural-HNT, Crystalline-HNT and Amorphous-HNT, and exposed to distilled water at 70°C for up to 336 h (14 days). As can be seen, the maximum moisture absorption rate and the stabilization time of the moisture absorption rate change with the crystallinity and content of HNT. These data can be used to calculate the equilibrium moisture content. The hygroscopicity characteristics in GFRP and BFRP nanocomposites were improved by 18.6%, 35.3%, 24% and 25.7% of 0.5 wt% Crystalline-HNT/GFRP, and 0.5, 1 and 3 wt% Amorphous-HNT/GFRP, respectively, and 7.3%, 32.4%, 17.3% and 15.1% of 0.5 wt% Crystalline-HNT/BFRP, and 0.5, 1 and 3 wt% Amorphous-HNT/BFRP, respectively, compared to Neat GFRP and Neat BFRP. At 0.5 wt% Amorphous-HNT, both GFRP and BFRP showed a water absorption resistance improvement of 30% more, and the stabilization time of the absorption rate of Amorphous-HNT/GFRP was 2.5 times longer than Neat GFRP. The results show the effect of HNTs in significantly reducing internal damage to the composite exposed to high-temperature water for a long time. Thus, depending on its crystallinity and content in the nanocomposites, HNTs can prevent the degradation of epoxy by

Table 1.3 Results of water absorption test (ASTM D5229) of the nanocomposites in distilled water at 70°C according to crystallinity and content of HNT

Samples

Code	HNT amount (wt%)	Number of laminated layers	Maximum water absorption rate (%)	Stabilization time of absorption rate (h)
Neat GFRP	N/A	1	2.03	24
		2	1.43	36
		4	2.37	36
		6	1.38	36
		12	1.12	36
		Average	**1.67**	**34**
Crystalline-HNT/GFRP	0.5	1	1.78	43
		2	1.56	43
		4	1.40	84
		6	0.90	96
		12	1.18	84
		Average	**1.36**	**70**
	1	1	1.98	96
		2	2.41	96
		4	1.59	72
		6	2.50	240
		12	3.97	288
		Average	**2.49**	**158**
	3	1	2.53	31
		2	1.47	68
		4	2.05	89
		6	3.39	163
		12	0.87	96
		Average	**2.06**	**89**

Samples

Code	HNT amount (wt%)	Number of laminated layers	Maximum water absorption rate (%)	Stabilization time of absorption rate (h)
Neat BFRP	N/A	1	2.86	48
		2	1.51	36
		4	1.81	60
		6	1.66	108
		12	1.10	84
		Average	**1.79**	**67**
Crystalline-HNT/BFRP	0.5	1	1.94	67
		2	2.15	91
		4	1.23	60
		6	1.56	96
		12	1.40	192
		Average	**1.66**	**101**
	1	1	1.84	12
		2	2.73	48
		4	1.36	36
		6	2.04	12
		12	1.39	144
		Average	**1.87**	**50**
	3	1	3.32	117
		2	1.61	20
		4	3.35	86
		6	2.78	96
		12	2.58	166
		Average	**2.73**	**97**

Table 1.3   (*Continued*)

Samples					Samples				
Code	HNT amount (wt%)	Number of laminated layers	Maximum water absorption rate (%)	Stabilization time of absorption rate (h)	Code	HNT amount (wt%)	Number of laminated layers	Maximum water absorption rate (%)	Stabilization time of absorption rate (h)
Amorphous-HNT/GFRP	0.5	1	2.34	48	Amorphous-HNT/BFRP	0.5	1	2.33	24
		2	0.96	120			2	1.08	96
		4	0.73	72			4	0.72	12
		6	0.69	120			6	0.97	96
		12	0.66	72			12	0.94	72
		Average	**1.08**	**86**			Average	**1.21**	**60**
	1	1	1.77	216		1	1	1.68	24
		2	1.17	24			2	1.56	72
		4	1.14	96			4	1.42	36
		6	1.28	120			6	1.40	36
		12	0.97	96			12	1.33	96
		Average	**1.27**	**110**			Average	**1.48**	**53**
	3	1	1.82	24		3	1	2.30	36
		2	0.96	48			2	1.11	48
		4	1.18	96			4	1.41	24
		6	1.10	120			6	1.37	120
		12	1.13	96			12	1.42	96
		Average	**1.24**	**77**			Average	**1.52**	**65**

water by lowering the maximum moisture absorption rate within a specific time range and by extending the time to reach the equilibrium moisture content. Thus, the fatigue life of structural materials can be increased in extreme environments, such as water, by adding HNTs, which indicates the great application potential of HNTs.

Figure 1.12 depicts the change in the water absorption rate with the number of laminated layers. Both single-layered GFRP and BFRP are vulnerable to moisture and showed a stable tendency to water (high water absorption resistance) in Amorphous-HNT as compared to Crystalline-HNT. Nanocomposites with a 1 wt% or higher Crystalline-HNT content were partially vulnerable to water due to particle agglomeration. Agglomerates of a certain size or larger remain in the powder form and create unimpregnated areas that do not form chemical bonds with epoxy. Thus, they serve as a passage through which water rapidly penetrates the composites due to the micro-damages generated on the surface of the epoxy nanocomposite. Eventually, the water absorption rate of the nanocomposite exceeds that of pure epoxy, which causes degradation. Therefore, 0.5 wt% Amorphous-HNT in GFRP and BFRP, except in the case of single-layered composites in which interface formation is impossible, exhibited excellent dispersibility and chemical bonding with epoxy and an excellent bridging effect for both glass and basalt fibers. Thus, Amorphous-HNT improved the resistance to water absorption stably regardless of the number of layers. Moreover, unlike Crystalline-HNT, Amorphous-HNT is not prone to aggregation and thus exhibits structural stability.

Figure 1.13(a) depicts the change in the water absorption rate of four-layered laminated GFRP and BFRP nanocomposites against the water immersion time. As mentioned earlier, HNT significantly improves moisture absorption regardless of its crystallinity. Furthermore, HNT content is an important factor to achieve a uniform dispersion in epoxy, and the need for surface modification of HNT will be determined based on its content in the composite to ensure its structural stability.

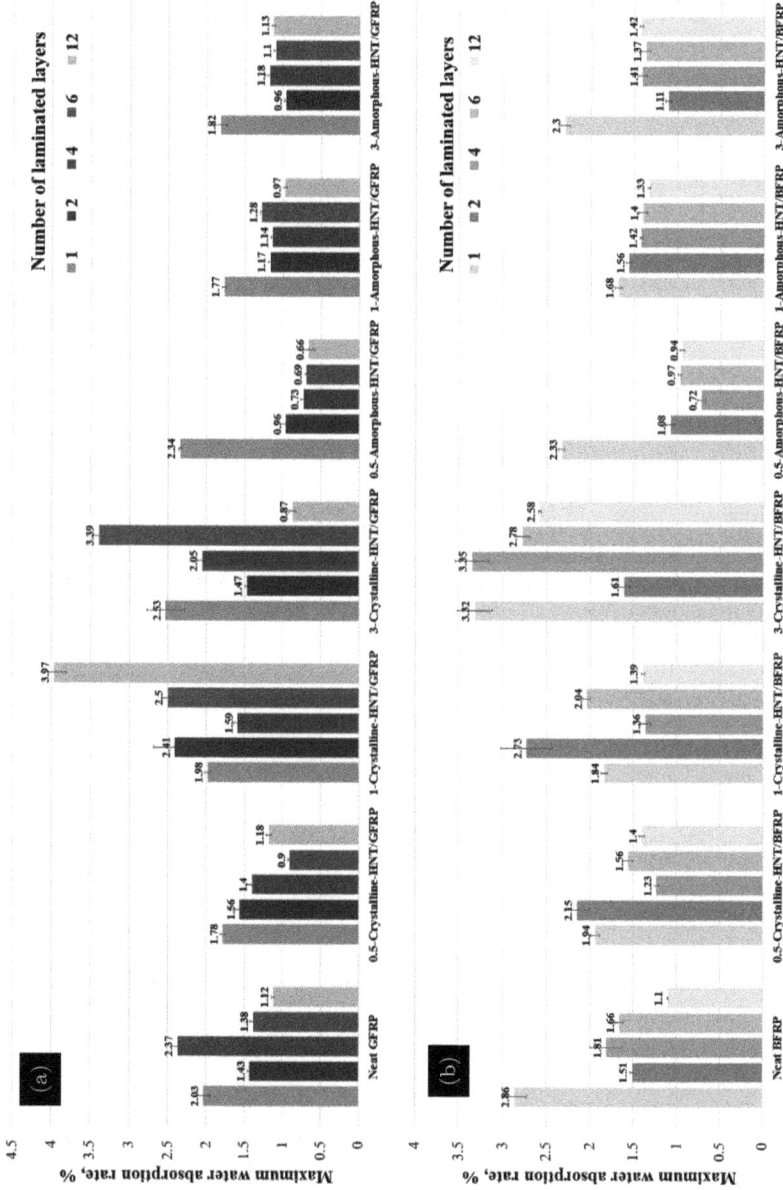

Fig. 1.12 Moisture absorption rate of HNT reinforced GFRP (a) and BFRP (b) composites according to the number of laminated layers.

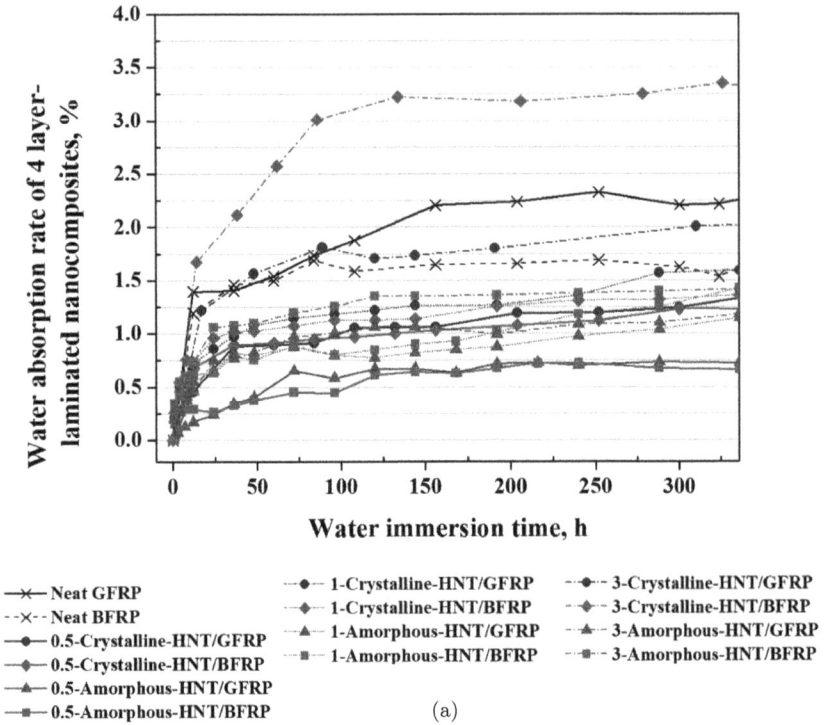

(a)

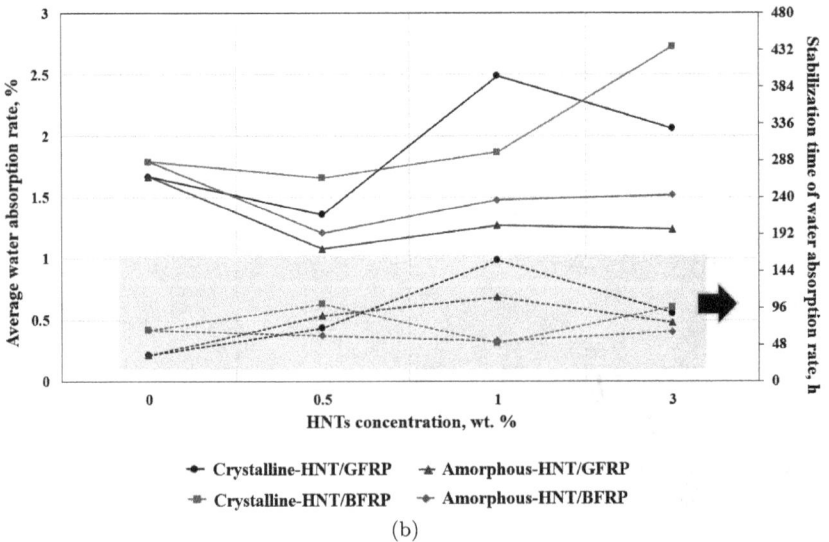

(b)

Fig. 1.13 (*Continued*)

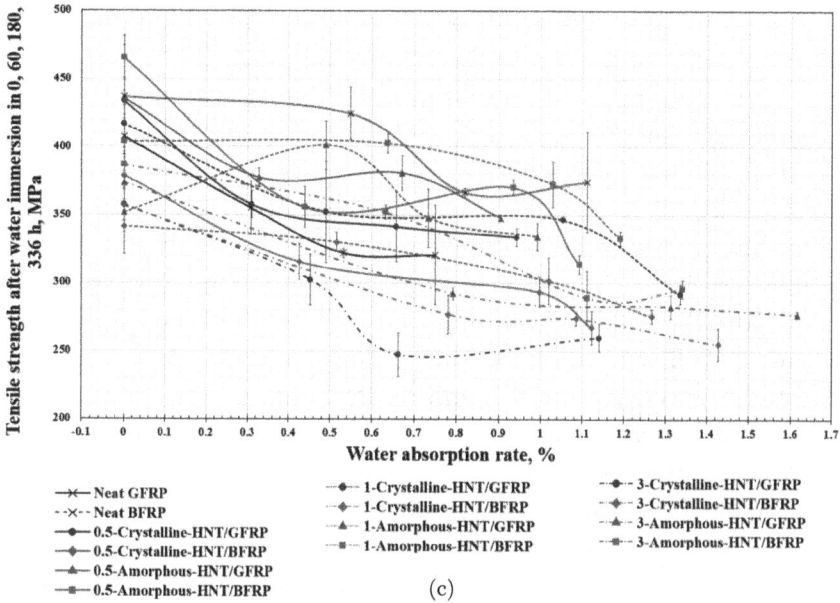

Fig. 1.13 Correlation analysis between moisture environment degradation-mechanical strength of HNT reinforced GFRP and BFRP composites. (a) Water absorption behavior according to the crystallinity and content of HNT in four-layered laminated CFRP and BFRP nanocomposites. (b) The effect of HNT crystallinity on water absorption behavior according to HNT content. (c) Changes in the tensile strength and the water absorption rate of HNT reinforced GFRP and BFRP composites with increasing water absorption time.

Figure 1.13(b) depicts the effect of HNT crystallinity on the water absorption behavior of the nanocomposites with various HNT contents. The water absorption rate and the stabilization time of the water absorption rate of Crystalline-HNT/GFRP and Crystalline-HNT/BFRP did not change significantly with HNT content. However, at a content of 1 wt% or more, the utilization of HNTs will be limited due to insufficient dispersion and structural stability. In contrast, Amorphous-HNT/GFRP and Amorphous-HNT/BFRP exhibited lower water absorption rates and the stabilization time of water absorption rate compared with those of Neat GFRP and Neat BFRP, respectively. Thus, the structural stability of Amorphous-HNT promotes its bonding with the epoxy matrix.

Figure 1.13(c) depicts the effect of moisture absorption on the tensile strength of the nanocomposites immersed in high-temperature water for a long time. The tensile test (ASTM D3039) was performed after 60, 180 and 336 h of immersion in water and the relationship between the water absorption rate and the tensile strength was determined. The tensile strength of Amorphous-HNT was higher than that of Crystalline-HNT above a certain water absorption rate at all conditions. The tensile strength of 0.5 and 1 wt% Amorphous-HNT/GFRP and 0.5 wt% Amorphous-HNT/BFRP increased at 180, 60 and 180 h, respectively. Since Amorphous-HNT is superior to Crystalline-HNT in terms of interfacial bonding strength and dispersibility, the rate of decrease in tensile strength of the Amorphous-HNT-based nanocomposites was low. The tensile strength of 1 wt% Amorphous-HNT/GFRP was higher than that of the nanocomposite not immersed in water. This increase in tensile strength is because of the increase in resistance to external loads due to the rapid diffusion of water into the interior of the powder-formed Amorphous-HNT not combined with epoxy. Particle clusters inside composite laminates act like pores and easily break or damage under external loads. However, when the empty space is filled with a relatively dense liquid, the external load is absorbed by the liquid, which temporarily enhances the reinforcing effect. Further, while the water absorption rate of the Amorphous-HNT-based nanocomposite was greater than that of the 1 wt% Crystalline-HNT nanocomposite, their reinforcement rate and the rate of decrease in tensile strength were similar. In conclusion, HNT increases ductility and improves the fracture resistance by increasing yield. In the case of Amorphous-HNT, aggregation increased with increasing content; however, this did not directly damage the glass, basalt fibers or the epoxy matrix. Instead, by binding the water molecules, Amorphous-HNT improved the mechanical properties, and consequently slowed the activation of epoxy in water.

In the case of GFRP nanocomposites, the mechanical properties and water resistance were significantly affected by the crystallinity and content of HNT, whereas in the BFRP nanocomposites, damage was caused by a variation in the content of HNT depending on its size and shape. Crystalline-HNT and Amorphous-HNT increase the ductility and delay the occurrence of yield, thereby reducing deformation under external loads and increasing resistance to damage. Amorphous-HNT facilitated the penetration and diffusion of moisture, while Crystalline-HNT was greatly influenced by the degree of dispersion in epoxy (homogeneous dispersion or not). Therefore, particle aggregates ultimately inhibited the physical properties of the material, such as a stress concentration part. In contrast, Amorphous-HNT formed clusters and combined with water, thereby trapping the water inside the cluster, which prevented the water from directly damaging the material. Moreover, it increased the tensile modulus and improved the stiffness.

Based on the water absorption test results, the interfacial bonding of the reinforcing fibers and the epoxy matrix and the effect of HNT on the defect damage caused by water were analyzed by scanning electron microscopy (SEM). In the water absorption test, the samples were immersed in distilled water for up to 336 h. During this time, neat epoxy was deteriorated by water and the bonding between the epoxy matrix and fibers weakened, which can lead to fracture under a small external load. Additionally, the condition facilitates the formation of micro-cracks, causing irreversible damage due to which strength recovery is impossible. The microscopic analysis was conducted to determine the factors responsible for fiber damage due to the delamination of the HNT/epoxy clusters degraded by water and the deterioration of interfacial strength due to HNT aggregation. Figure 1.14(a) depicts the interface of Neat GFRP without HNT before and after water absorption. Before water absorption, the glass fibers are bonded to epoxy, whereas after water absorption, most of the epoxy is delaminated and the residual epoxy clusters are removed. The epoxy expanded with water, peeled off from the fiber surface with a coarse fracture, and gradually fractured over a long period. The expanded epoxy provided a diffusion pathway for moisture from the glass fiber–epoxy interface to the interior, which eventually led to fiber breakage when the moisture content of the glass fiber reached saturation. As evidenced from Fig. 1.14(b), Crystalline-HNT improved the interfacial bonding. Water softened

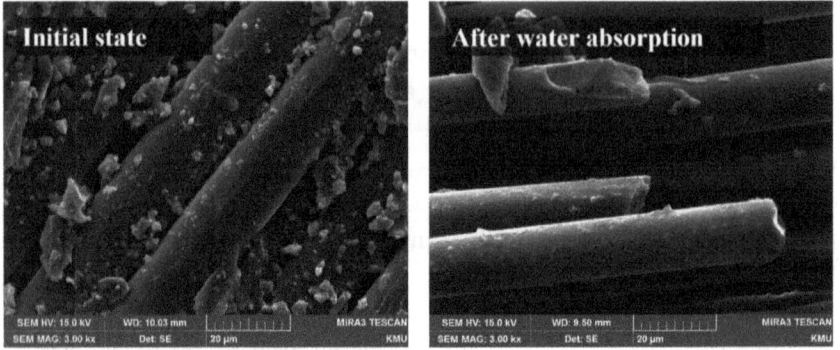

(a) Neat GFRP

(b) Crystalline-HNT/GFRP

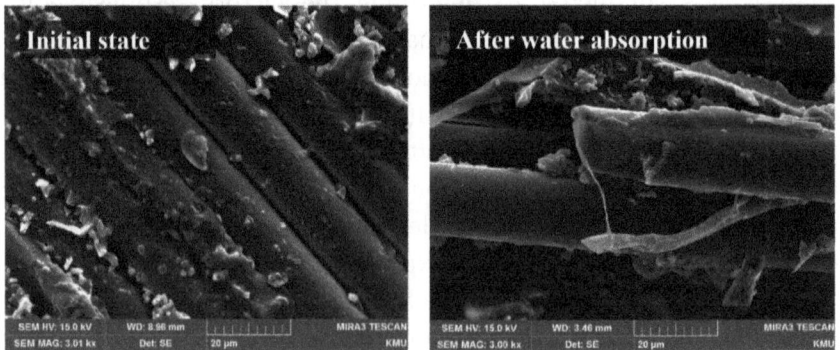

(c) Amorphous-HNT/GFRP

Fig. 1.14 SEM observation at the interface of the GFRP nanocomposite laminates before (left) and after (right) water immersion at 70°C.

the initial epoxy clusters with a sharp shape, and the fiber surface swelled. This meant that some were removed by moisture, and some remained on the surface as a fully bonded form between Crystalline-HNT and epoxy, supporting the glass fibers, thus minimizing the direct water exposure effects. In Fig. 1.14(c), Amorphous-HNT was combined with a large area of the fiber surface. A small amount of dull epoxy fragments, residue generated by the delamination of epoxy on the glass fiber surface after immersion in water, was also observed. Unlike in the case of Crystalline-HNT, interfacial delamination due to water saturation is the main cause of damage to Amorphous-HNT rather than fiber breakage due to water diffusion into the fiber.

Figure 1.15(a) depicts the SEM image of a Neat BFRP. As can be seen, the basalt fiber-epoxy interface is smooth; however, after prolonged immersion in water, epoxy delamination and fiber breakage occurred, which resulted in numerous sharp cuts and dents on the fiber surface. Thus, while the basalt fibers and epoxy matrix were strongly bonded, water directly affected the fibers and deteriorated the strength. As shown in Fig. 1.15(b), Crystalline-HNT/BFRP has a smoother initial fiber surface compared to that of Crystalline-HNT/GFRP (Fig. 1.14(b)). However, water absorption led to severe internal damage in the form of fiber breakage. As shown in Fig. 1.15(c), in the case of Amorphous-HNT, the basalt fibers and epoxy matrix are uniformly bonded, while in the case of Crystalline-HNT, some clusters are observed. Additionally, long scratches and wrinkles can be seen. Immersion in water resulted in a rough fiber surface with sharp cuts, which led to instantaneous fiber breakage.

For structures that are used in harsh environments such as a water environment, it is necessary to predict the durability of the constituent materials. Since composite materials generally have high corrosion resistance, it is difficult to determine their durability in appearance without any damage (destruction evaluation). However, a better understanding of the destruction mechanism of the composite material under certain conditions, such as temporary impact failure or fatigue failure, will allow us to determine the factors that can cause fatal damage even during non-destructive evaluation. The incorporation of a small amount of HNTs is not only economically

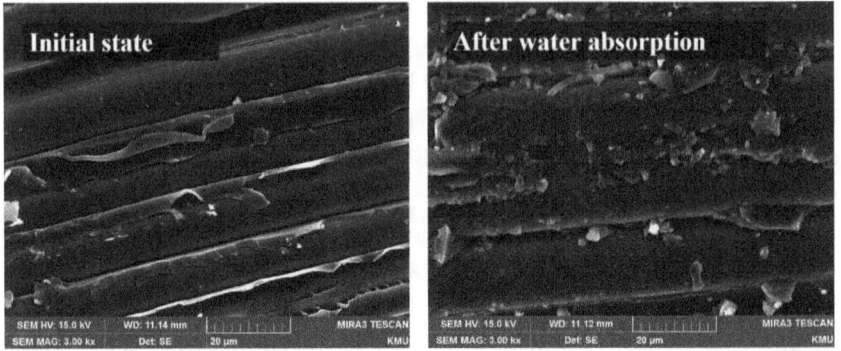

(a) Neat BFRP

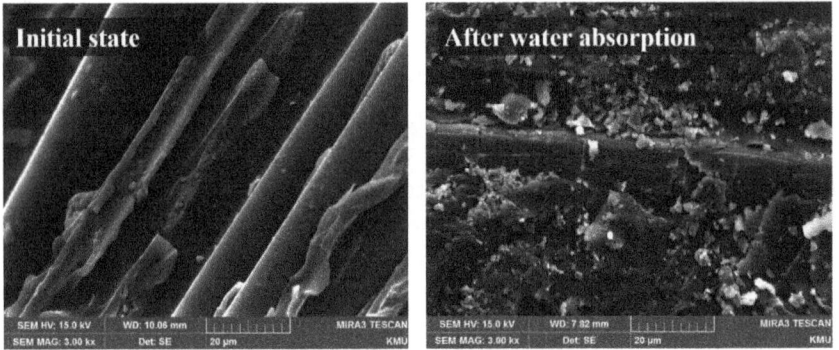

(b) Crystalline-HNT/BFRP

(c) Amorphous-HNT/BFRP

Fig. 1.15 SEM observation at the interface of the BFRP nanocomposite laminates before (left) and after (right) water immersion at 70°C.

(a) Crystalline-HNT/GFRP

(b) Crystalline-HNT/BFRP

(c) Amorphous-HNT/GFRP

(d) Amorphous-HNT/BFRP

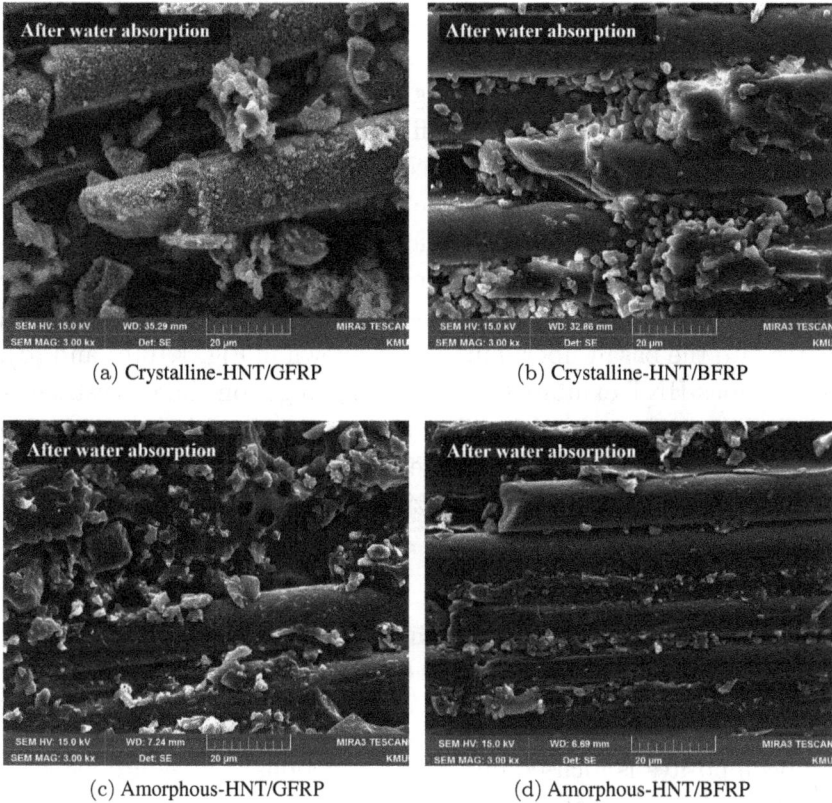

Fig. 1.16 SEM observation at the interface of HNT reinforced GFRP and BFRP nanocomposite laminates after water immersion at $70°C$ up to 700 h.

viable but also technologically advantageous as it enhances the water absorption resistance and improves structural stability and durability by minimizing internal defects. Thus, the effect of HNT crystallinity was further analyzed to determine the internal change caused by prolonged immersion in water leading to water contents higher than the saturation amount.[36]

Figure 1.16 depicts the SEM image on the interface after 700 h of water immersion. As depicted in Fig. 1.16(a), Crystalline-HNT/GFRP exhibits a damaged surface with fiber breakage. Residual Crystalline-HNT/epoxy clusters surrounding the glass fiber surface expanded because HNT trapped the water that penetrated the composite and prevented its discharge. That is, Crystalline-HNT absorbs water directly by bonding around the epoxy and prevents its deterioration by moisture. During this process, the supersaturated water partially diffuses into the fiber, and the epoxy, which is not directly bonded to Crystalline-HNT, is removed by the adsorption and desorption of water, and the minimally Crystalline-HNT and epoxy clusters of the fully bonded form remains. In contrast, as shown in Fig. 1.16(b), Crystalline-HNT caused delamination of epoxy over the entire surface of the basalt fiber due to penetration of absorbed water into the basalt fiber. Further, as shown in Fig. 1.16(c) and (d), Amorphous-HNT exhibits less particle aggregation than Crystalline-HNT. The uniform particle dispersion and the stable bonding with both fiber types and the epoxy prevented any serious damage to the interior of the fiber, thus minimizing the severe damage caused by water.

1.4   Effects of HNTs on Mechanical Strength and Flammability under Carbonization[30,37]

Research and development in the field of nanomaterials and nanocomposites is focused on specific functions as they are low-cost and high-risk materials requiring a complex manufacturing process. The application of HNTs in large structural components is increasing not only because of its excellent dispersibility in polymer matrices and strong bonding with polymers but also because of the reinforcing effect, hygroscopicity, eco-friendliness and excellent flammability. In particular, the chemical composition of HNT makes it an excellent heat-resistant material. Thus, the flammability of HNTs was analyzed and the effect of its crystallinity on its flammability was determined.

As previously mentioned, the interfacial interaction between epoxy and HNTs is important to maximize the functionality of HNT. Polymers, such as epoxy, have a high mobility at high temperatures below the curing temperature, which can be controlled by modulating the content and morphological structure of HNTs. HNTs improve the mechanical properties and moisture absorption resistance of

Table 1.4 Flammability characteristics of HNT/GFRP and HNT/BFRP laminates according to the crystallinity of HNTs

| Samples* | Fiber weight fraction (wt%) | UL 94 5V test (ASTM D5048) | | | LOI (vol. % $O_2 \pm 2\sigma$) (ASTM D2863) | |
		Carbonized area (cm^2)	Level	Ignition	Extinguish	
Crystalline-HNT/GFRP**	60.62	34.0	5VA	20.77 ± 0.1	20.25 ± 0.1	
Amorphous-HNT/GFRP**	64.23	35.2	5VA	23.18 ± 0.4	22.3 ± 0.1	
Crystalline-HNT/BFRP**	58.67	17.6	5VA	22.81 ± 0.5	21.71 ± 0.2	
Amorphous-HNT/BFRP**	58.19	27.5	5VA	23.11 ± 0.4	21.8 ± 0.2	

Notes: *All samples included 1wt% HNT. **Plain fabric supplied by GM Composites Co., Republic of Korea.

nanocomposites and are expected to play an important role in improving the flame retardancy by increasing the thermal stability. Thus, HNT is incorporated to reduce the CTE and form a physical barrier during combustion to prevent heat transfer. This will expand the application of polymer-based HNT reinforced nanocomposites.

The data presented in Table 1.4 show the effect of HNT crystallinity on its flammability for GFRP and BFRP nanocomposites containing 1 wt% HNT, as determined by the UL 94 5V flammability test (ASTM D5048) and the limiting oxygen index (LOI) (ASTM D2863). All the samples used in the test had a fiber weight fraction of approximately 60 wt%. In the UL 94 5V test, the surface of the FRP laminate was ignited with a torch and the burning surface was observed. All the samples exhibited excellent flammability with a rating of 5VA, and neither cracks nor flame penetration was observed on the surfaces of the FRP laminates. The only difference was in the size of the carbonized area. The carbonized area of Crystalline-HNT/GFRP was over 48% larger than that of Crystalline-HNT/BFRP, whereas the carbonized area of Amorphous-HNT/GFRP was over 22% larger than that of Amorphous-HNT/BFRP.

The smaller carbonized area of the HNT/BFRP laminates indicates the excellent bonding between the HNTs and the basalt fibers and epoxy matrix, which improved their flammability. In the case of instantaneous heat (flame), Crystalline-HNT and Amorphous-HNT in BFRP delayed heat transfer by forming a physical barrier on the carbide surface owing to their similar chemical composition.

As indicated by the LOI, Crystalline-HNT/GFRP caught flame instantaneously in an air atmosphere. However, Amorphous-HNT/GFRP, Crystalline-HNT/BFRP and Amorphous-HNT/BFRP burned slowly in an air atmosphere. Thus, even considering the different types of fiber reinforcements, Amorphous-HNT resisted char generation better than Crystalline-HNT did. The carbonized areas acted as physical barriers on the surface. Further, the ILSS and flexural strength were evaluated to determine the effect of carbonization on mechanical strength. Figure 1.17(a) shows the variation in the interfacial bond strength with heat treatment temperature that dominates the crystallinity of HNT. It is noteworthy that regardless of its crystallinity, HNT significantly improved the interfacial bonding strength of the BFRP. The structure of Amorphous-HNT decreased the interfacial bonding strength of the GFRP laminate but increased the interfacial bonding strength of the BFRP laminate to more than twice that of Neat BFRP. The epoxy matrix is the same in all the FRP laminates, however, the difference in type and chemical composition of the fiber reinforcements demonstrated the importance of HNT in the cross-linking of the fiber reinforcement and epoxy matrix. This is because epoxy and HNT form a colloidal dispersion and are thus involved in the curing reaction.

As shown in Fig. 1.17(b), before combustion, HNTs exhibit a remarkable reinforcing effect of the flexural strength for BFRP but not for GFRP. Crystalline-HNT and Amorphous-HNT-based BFRP nanocomposites exhibited an excellent strengthening effect compared with the Neat BFRP laminate, which demonstrated a low flexural strength. Additionally, the crystallinity of HNT significantly affected the flexural modulus. However, after carbonization, the flexural strengths and flexural moduli of HNT-containing GFRP and BFRP were similar regardless of the crystallinity of HNT. This is because

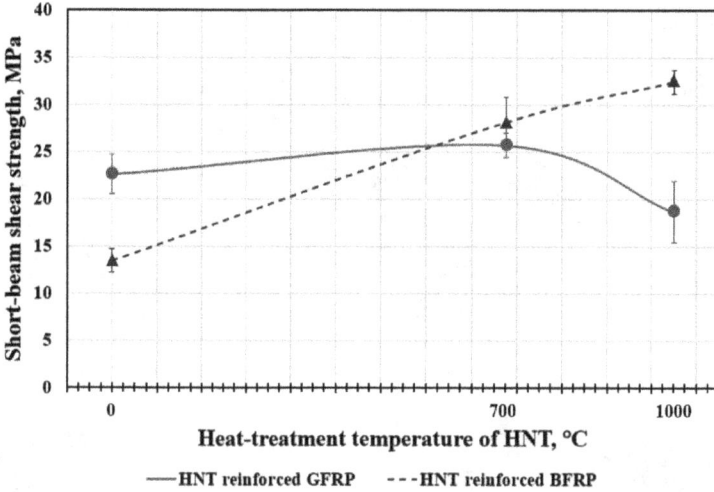

(a) Inter-laminar shear strength

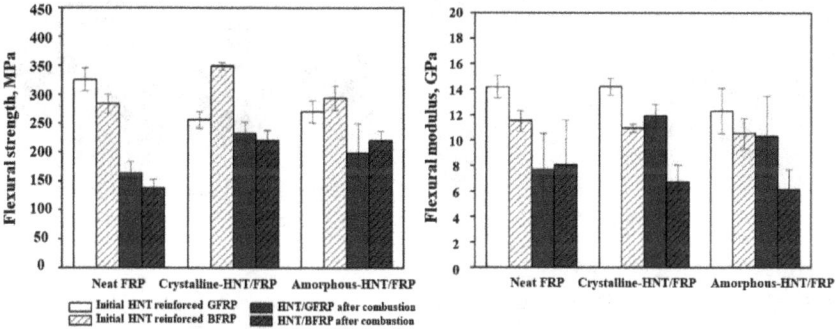

(b) Flexural strength (left) and flexural modulus (right)

Fig. 1.17 Results of mechanical properties according to the crystallinity of HNT before and after the combustion test.

of the structural transformation of HNT in the epoxy matrix upon exposure to a high-temperature flame.

Further, Fig. 1.18 depicts the effect of carbonization on the mechanical strength of the nanocomposites. For HNT/BFRP, the reinforcing effect of HNTs was insufficient due to the superimposition and localization of the carbonized and charred areas. Thus, the rate of decrease in strength of Crystalline-HNT/BFRP was larger

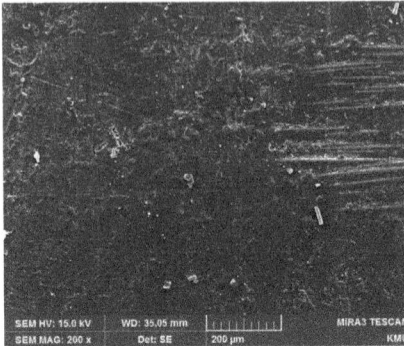

(a) Initial surface of GFRP laminate

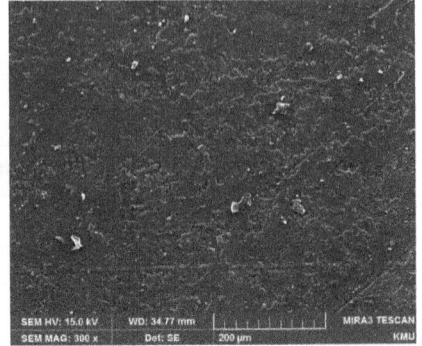

(b) Initial surface of BFRP laminate

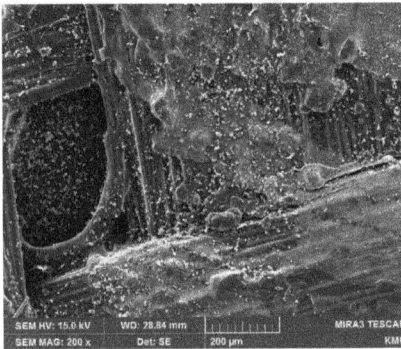

(c) Crystalline-HNT/GFRP after carbonization

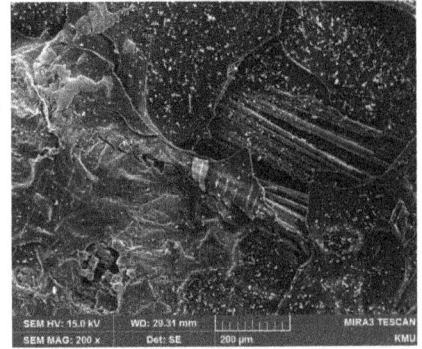

(d) Crystalline-HNT/BFRP after carbonization

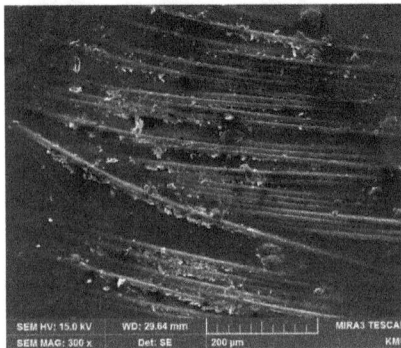

(e) Amorphous-HNT/GFRP after carbonization

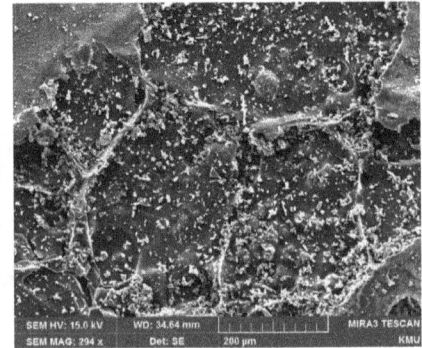

(f) Amorphous-HNT/BFRP after carbonization

Fig. 1.18 SEM surface observation of HNT/GFRP and HNT/BFRP laminates according to the crystallinity of HNTs after a UL 94 5V test through SEM.

than that of Crystalline-HNT/GFRP. In contrast, since HNT/GFRP has a large carbonized area, carbides are uniformly generated and remain on the surface, and the stress concentration under an external load is small. In the case of Crystalline-HNT/GFRP, which caught flame over a large area, the glass fibers were extensively damaged due to prolonged exposure to a high-temperature flame. This is possibly the reason for the significant strength reduction due to the carbonization reaction, even though the ILSS of Crystalline-HNT/GFRP is the same as that of Crystalline-HNT/BFRP. Furthermore, the effects of Amorphous-HNT and Crystalline-HNT on the initial mechanical/interfacial strength of GFRP were the same; however, they exhibited a physical barrier effect after carbonization. Generally, the mechanical strength of composite laminates depends on the type of fiber reinforcement. However, HNT, regardless of its crystallinity, improved the heat stability of the nanocomposite through the generation of carbides by carbonization that acted as a physical barrier. In conclusion, regardless of its crystallinity, HNT acted as a surface barrier in GFRP and BFRP laminates, thereby preserving the inherent mechanical strength owing to its excellent flammability. Thus, the incorporation of HNT will allow the excellent mechanical strength of the composite laminate when burned to be maintained.

1.5   Conclusions

To enhance the strength of multi-layered hybrid polymer composites using nanofillers, the following two conditions must be satisfied: the uniform dispersion of nanofiller in the matrix, and a stable interfacial bonding strength. The dispersibility of nanoparticles in the polymer matrix is important since polymers are viscous and thus exhibit heat-sensitive behavior, such as heat shrinkage, during the curing reaction; therefore, a strong bonding interaction between the polymer and nanoparticles is necessary. Non-uniform dispersion of nanoparticles in the polymer matrix is mainly because of the high viscosity of the polymer and the strong cohesion between the nanoparticles. Since the incorporation of only a small amount of nanofillers can impart excellent physical properties, the performance of the materials can be enhanced with a relatively simple manufacturing process through the physical/chemical/structural surface modification of nanofillers. A stable interfacial bonding strength can be achieved by enhancing the bonding strength between the polymer matrix and the fiber reinforcement by incorporating nanoparticles, which will enhance not only the structural stability but also the resistance to external environmental factors. That is, even if a uniform polymer-nanoparticle dispersion system is achieved, the possibility of the formation of micro-defects during the impregnation process cannot be excluded. Moreover, composite materials comprise two or more dissimilar materials that exhibit different characteristics; thus, even if durability is enhanced, the composite may be vulnerable to environmental factors such as water or heat, which will limit the application scope of the composite.

In this respect, HNT, a naturally occurring mineral with a variety of structures depending on its crystallinity, exhibits a great application potential owing to its excellent properties. In this chapter, the effects of crystallinity, water absorption and flammability of HNTs on the mechanical properties, such as tensile strength and ILSS of HNT-based fiber reinforced epoxy matrix nanocomposites are discussed. The thermal analysis of HNTs provided a better understanding of the evolution of the crystallinity of HNTs. The results suggested that a composite of HNT and basalt fibers could be used for the development of eco-friendly materials. Furthermore, the analysis of HNT-based composites comprising a thermosetting epoxy matrix and glass or basalt fiber reinforcements, which are representative structural composites, provides important insights for the application of these composites as high-performance industrial materials. In the future, the design of HNT-based polymer composites and improvements in their dispersibility through surface modification are expected to provide a better performance control through the combination of materials for application as high-performance engineering and structural materials.

Acknowledgment

This chapter is based on the research history of first author Soo-Jeong Park, and is a comprehensive compilation of her research achievements in the recent 7 years, published through World Scientific Publishing Company. The first author appreciates co-author Prof. Yun-Hae Kim's generous support and dedicated academic guidance for giving her the opportunity to write this chapter. Dr. Soo-Jeong Park would also like to thank her co-author Sung-Min Yoon, for his outstanding insight and help.

References

[1] Y. H. Kim, K. M. Moon, Y. D. Jo, S. Y. Bae, S. J. Sin, M. J. Kim and J. Y. Kim, *Adv. Sci. Lett.*, 4, 2011, 1455.

[2] S. M. Yoon, S. J. Park and Y. H. Kim, *Mod. Phys. Lett. B*, 33, 2019, 1940026.

[3] Y. H. Kim, S. J. Park, J. W. Lee and K. M. Moon, *Mod. Phys. Lett. B*, 29(06n07), 2015, 1540003.

[4] Y. H. Kim, A. N. Nakagaito and S. J. Park, Analysis Optimal Process Conditions and Mechanical Properties on Nanocomposites according to Structural Changes of Halloysite Nanotubes, 8th International Conference on Physical and Numerical Simulation of Material Processing, Seattle, Washington, 2016.

[5] S. H. Hwang, D. W. Kang and H. J. Kang, *Polym. Korea*, 40, 2016, 967–971.

[6] F. Fatemeh, E. Morteza and J. Ali, *Thermochim. Acta*, 568, 2013, 67–73.

[7] M. A. Kibria, M. R. Anisur, M. H. Mahfuz, R. Saidur and I. H. S. C. Metselaar, *Energy Convers. Manag.*, 95, 2015, 69–89.

[8] N. F. Attia, M. A. Hassan, M. A. Nour and K. E. Geckeler, *Polym. Int.*, 63, 2014, 1168–1173.

[9] C. R. Choe, *Compos. Res.*, 26, 2013, 147–154.

[10] S. C. Lee, J. H. Kim, G. S. Choi, Y. D. Jang, A. Amanov and Y. S. Pyun, *Tribol. Lubr.*, 31, 2015, 56–61.

[11] Sun Moon University, *A Development of Nano Crystalline Surface Modification Technology*, Asan, 2012.

[12] C. Li, J. Liu, X. Qu and Z. Yang, *J. Appl. Polym. Sci.*, 112, 2009, 2647–2655.

[13] S. M. Yoon and Y. H. Kim, *Int. J. Mod. Phys. B*, 32, 2018, 1840051.

[14] R. Talrega, *J. Strain Anal. Eng. Des.*, 24, 1989, 215.

[15] S. R. White *et al.*, *Nature*, 409, 2001, 794.

[16] C. Janssen, *Proc. 10th Int. Congr. Glass in Tokyo*, 10, 1974, 23.

[17] A. Grimaldi *et al.*, *Phys. Rev. Lett.*, 100, 2008, 165505.

[18] T. A. Plaisted, A. V. Amirkhizi and S. Nemat-Nasser, *Int. J. Fract.*, 141, 2006, 447.

[19] G. Pallares *et al.*, *Int. J. Fract.*, 156, 2009, 11.

[20] B. Szczpanik, P. Slomkiewicz, M. Garnuszek, K. Czech, D. Banas, A. Kubala-Kukus and I. Stabrawa, *J. Mol. Struct.*, 1084, 2015, 16–22.

[21] E. Horvath, J. Kristof, R. Frost, A. Redey, V. Vagvolgyi and T. Cseh, *J. Therm. Anal. Calorim.*, 71, 2003, 707–714.

[22] Y. Dong, B. Lisco, H. Wu, J. H. Koo and M. Krifa, *J. Appl. Polym. Sci.*, 132, 2015, 275–287.

[23] Z. Shu, Y. Chen, J. Zhou, T. Li, D. Yu and Y. Wang, *Appl. Clay Sci.*, 112, 2015, 17–24.

[24] G. Cavallaro, D. I. Donato, G. Lazzara and S. Milioto, *J. Phys. Chem. C*, 115, 2011, 20491–20498.

[25] P. Yuan, D. Tran and F. Annabi-Bergaya, *Appl. Clay Sci.*, 112, 2015, 75–93.

[26] C. Duce, S. V. Ciprioti, L. Ghezzi, V. Ierardi and M. R. Tine, *J. Therm. Anal. Calorim.*, 121, 2015, 1011–1019.

[27] E. Gasparini, S. C. Tarantino, P. Ghigna, M. P. Riccardi, E. I. Cedillo-Gonzalez, C. Siligardi and M. Zema, *Appl. Clay Sci.*, 80, 2013, 417–425.

[28] M. Liu, B. Guo, M. Du, X. Cai and D. Jia, *Nanotechnol.*, 18, 2007, 455703.

[29] C. Dionisi, N. Hanafy, C. Nobile, M. L. De Giorgi, R. Rinaldi, S. Casciaro and S. Leporatti, *IEEE Tran. Nanotechnol.*, 15, 2016, 720–724.

[30] S. J. Park, Enhancement of Dispersion Stability on the Polymer Nanocomposites with Clay Nanoparticles through Analysis of Moisture Absorption Behavior, Doctoral Thesis, Busan, Republic of Korea, 2019.

[31] S. J. Park, Changes of the Structural and Mechanical Properties on Nanocomposites based on Halloysite Nanotubes with Dispersion Optimization by Ultrasonication, Master Thesis, Tokushima, Japan, 2016.

[32] S. J. Park, Changes of the Structural and Mechanical Properties on Nanocomposites based on Halloysite Nanotubes with the Optimization of Dispersion by Ultrasonic Waves, Master Thesis, Busan, Republic of Korea, 2016.

[33] S. Kadi, S. Lellou, K. Marouf-Khelifa, J. Schott, I. Gener-Batonneau and A. Khelifa, *Micropor. Mesopor. Mat.*, 158, 2012, 47–54.

[34] K. Sohlerg, S. J. Pennycook and S. T. Pantelides, *J. Am. Chem. Soc.*, 121, 1999, 7493–7499.

[35] Y. H. Kim, S. J. Park, J. S. Choi, K. M. Moon and C. W. Bae, *Int. J. Mod. Phys. B*, 32, 2018, 1840070.

[36] S. J. Park, J. S. Choi, T. Yu, Z. Chen, Y. H. Kim and C. W. Bae, *Mod. Phys. Lett. B*, 33, 2019, 1940020.

[37] Y. H. Kim, J. S. Choi and S. J. Park, *Mod. Phys. Lett. B*, 33(4n15), 2019, 1940021.

Chapter 2

Processing of Hierarchical-Distributed Halloysite Nanotube (HNT) Reinforced Composites by Electrophoretic Deposition

Tianyu Yu[*,‡] and Yun-Hae Kim[†,§]

*Major of Materials Engineering,
Department of Marine Equipment Engineering,
Korea Maritime and Ocean University,
Busan, Republic of Korea
†Department of Ocean Advanced Materials
Convergence Engineering,
Korea Maritime and Ocean University,
Busan, Republic of Korea
‡yutianyukmou@gmail.com
§yunheak@kmou.ac.kr

This chapter introduces an innovative electrophoretic deposition (EPD) method to achieve homogeneous and hierarchical distribution of Halloysite nanotubes (HNTs). The hierarchical distributed nanoadditives have the potentials to dramatically enhance the mechanical properties of the fiber reinforced polymers (FRPs), especially matrix-dominated properties like interfacial and interlaminar strength. The multi-step multi-scale mechanical modeling is performed based on the Eshelby–Mori–Tanaka model to characterize the stiffness property of the hierarchically structured FRPs. The stability of suspension for EPD is characterized using the total interaction energy method. The kinetic of EPD is investigated to achieve a controllable deposition. The feasibility of depositing HNTs by EPD method has been discussed from the aspects of suspension stability, intertube interaction, zeta potential and hydrophobicity. Conducting EPD on non-conductive fabric and fiber treatment is also discussed. The last part introduces the synergetic effect and mechanism of surfactant and hybrid deposition.

2.1 Introduction

Fiber reinforced polymers (FRPs) are now extensively applied to various industries due to their features like high specific strength in conjunction with high stiffness, toughness, fatigue and oxidation resistance. The strength elevation is, however, affected with applied load transiting from matrix to fibers and interfacial bonding between fiber-matrix. The interface is defined as the area with a certain thickness formed by the common boundary between the fiber and the matrix, which can maintain contact and connection while carrying out load transfer between the fiber and the matrix.[1] The relatively weak interfacial and interlaminar strength is the main factor that limits the overall properties of laminated FRPs, which has been an essential focus of study over decades.[2,3] The weak interfacial and interlaminar property of FRPs make them still susceptible to unpredictable and catastrophic failure, which may cause severe problems during service.[4] The research of the composite interface is still at the semi-empirical and semi-quantitative level, and the understanding of composite interface needs to be developed. With the development of nanotechnology, adding nanomaterials into FRPs have been found to have abilities to effectively enhance the comprehensive properties since the last few decades. Carbon nanotubes (CNTs) are one of the most representative nanomaterials featured with extremely high specific mechanical properties and aspect ratio. However, CNTs are high priced and cannot be easily afforded. Moreover, toxic components of ingredients used in recipes and pollutants during CNT manufacture were reported, which may have potential environmental impact.[5]

Halloysite nanotubes (HNTs), with the chemical formula $Al_2Si_2O_5(OH)_4 \cdot 2H_2O$, are now emerging as trendsetters in green nanotechnology. HNTs, which are naturally formed over millions of years, are entirely harmless to human beings and environments.[6] As an alternative of synthetical nanotubes, HNT is a fine clay mineral that consists of hollow tubular-shaped particles with a multi-layered wall structure. The unique hollow tubular structures are formed as a result of strain caused by lattice mismatch between adjacent

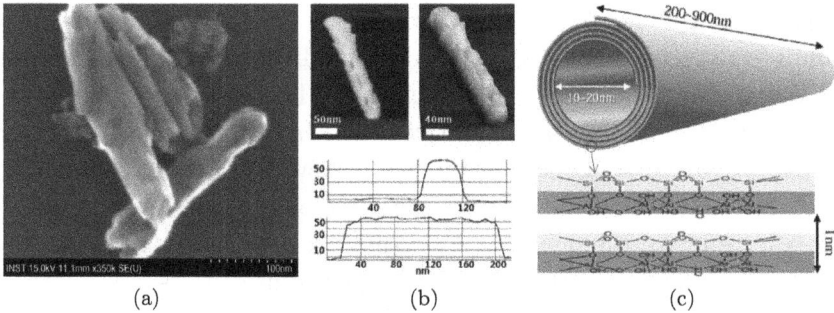

Fig. 2.1 The schematic with SEM and AFM images of HNT.[10]

silicon dioxide and aluminum oxide layers. The lengths of HNTs typically range from 0.5 to 3.0 μm, with exterior diameters ranging from 30 to 100 nm and internal diameters ranging from 15 to 30 nm. The schematic with scanning electron microscope (SEM) and atomic force microscopy (AFM) images of HNT is shown in Fig. 2.1. With their high aspect ratio, functionalities and reasonable price, HNTs may potentially act as an alternative to CNTs of the reinforcing nanoadditives for modifying fiber or matrix.[7] The remarkable properties such as impact resistance, flame resistance, high strength and modulus make HNTs reasonable to be employed to enhance the mechanical properties of FRPs. Ye *et al.* studied HNTs incorporated carbon fiber reinforced plastics (CFRPs) and found 5 wt% HNTs can improve 25% of interlaminar shear strength (ILSS), 37% of mode II fracture toughness and almost doubled mode II fracture toughness.[7] Prashantha *et al.* prepared HNTs/PA6 nanocomposites and found the enhancement of strength, modulus and dynamic mechanical properties like storage modulus and loss factor.[8] Yu *et al.* prepared HNTs/PLA nanocomposites and found the enhancement of mechanical and thermal properties associating with an accelerated cold crystallization of polylactic acid (PLA).[9]

To incorporate the nanoadditives into FRPs, the most conventional method is to disperse the nanoadditives into resin in a straigthforward way, with homogeneous processing like magnetic stirring or ultrasonic agitation. The well-dispersed resins are then used in FRPs fabrication step. By dispersing nanoadditives into fluidic epoxy, in an ideal state, the nanoadditives would be dispersed uniformly and randomly in the matrix. The matrix thus can be toughened by adding nanoadditives, which can be mathematically explained based on Eshelby's theory.[11] However, due to the random dispersion in the bulk matrix, the nanoadditives have minimal distribution to toughen the fiber-matrix interfacial regions. An innovative concept of the "hierarchical structure" of nano reinforced FRPs proposed that the nanoadditives were mainly distributed in the fiber–matrix interfacial region, rather than dispersing randomly in the matrix. A schematic of the hierarchical structure is illustrated in Fig. 2.2. The hierarchical dispersion of nanoadditives is expected to generate extreme mechanical and loading transport properties as well as possess multifunctional properties for applications in energy harvesting, capacitors or flexible batteries.[12] The primary benefit of hierarchical structures in laminated FRPs is to improve the matrix-dominated properties and through-thickness properties, notably the

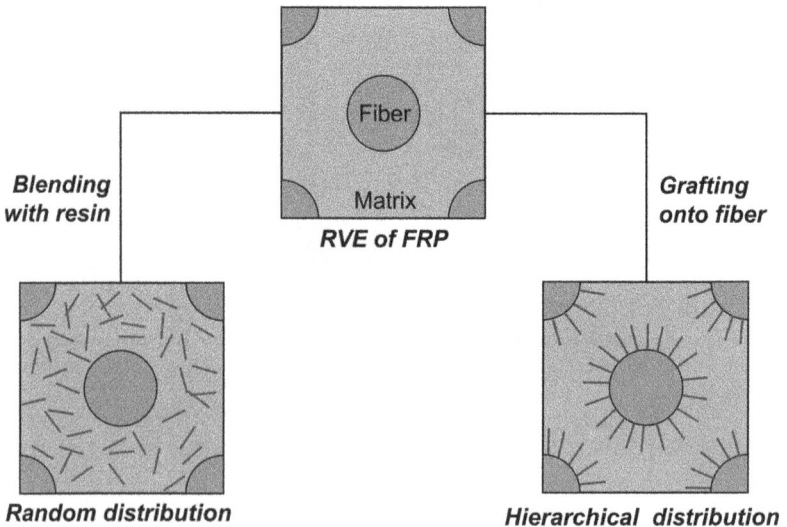

Fig. 2.2 Schematic of conventional FRP composites and hierarchical polymer composites.[16]

interfacial shear strength (IFSS), ILSS, transverse tensile strength and the resistance to intra and interlaminar failures, without compromising the in-plane performance.[13-15] The hierarchical-distributed nanoadditives contribute to a rougher fiber surface which interlocks the resin matrix by bridging mechanism. Moreover, the nanoadditives act as obstacles in the interfacial region, which can effectively impede the crack propagation by deflecting and traveling of the cracks. How hierarchical structures toughen the interfacial region is illustrated in Fig. 2.3.

There are several approaches to manufacture hierarchical-structured fabrics like sizing, coating, chemical grafting, selective placement, electrophoretic deposition (EPD), etc. Sizing is mostly common for commercial fibers during the synthesis steps, and the hierarchical structure can be realized by adding nanoadditives into the sizing agents. Coating is the simplest way to create the hierarchical structures just by dipping the dry fabrics into a suspension of nanoadditives, which may result in the inhomogeneous distribution of nanoadditives. To overcome this problem, the EPD method was investigated, which is believed to generate homogeneous and uniform deposition. This process was discovered by Bose in a liquid-siphon experiment during the 1740s and has been known

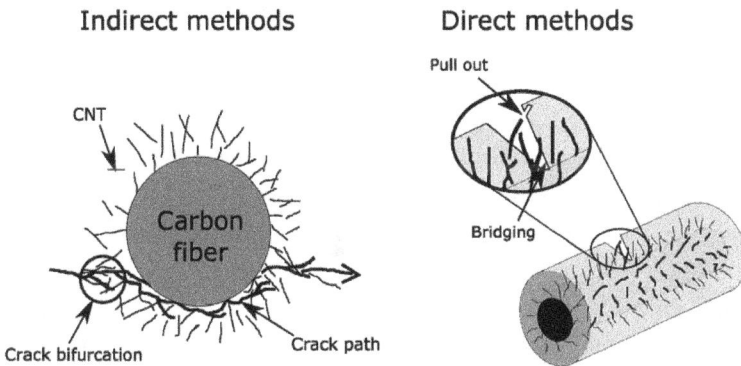

Fig. 2.3 Schematic of crack bifurcation, deflection mechanisms owing to the hierarchical nanodistribution.[12]

since Ruess observed the movement of clay particles in water under an electrical field.[17, 18] During EPD process, the charged nanoadditives are moved to the fabric having the opposite charge due to the electric field. Both alternating current (AC) and direct current (DC) electrical fields have been applied in EPD process, although DC fields are more common.[19] The charging of nanoadditives can be realized by using surfactant or modifications, and the zeta potential is always employed to characterize the changing degree and the mobility of nanoadditives. The EPD process can take place in a stable suspension of an aqueous solution or other organic solvents, while organic solvents always require more substantial field strength to realize the deposition. EPD technique's features of being conducive to highly versatile applications, low apparatus and equipment requirement, short processing time, cost-effectiveness, facile modification, desirable dense packing of particles in the final products, high quality of the microstructures produced, feasibility of geometrically complicated shapes and simple control, make EPD extensively applied for both research and industry purposes. The applications of EPD in the field of industry are, but not limited to, solar cells, batteries and electrochemical capacitors, solid oxide fuel cells and corrosion-resistant coatings. The schematic of nanoadditives dispersion, charging, electrophoresis and deposition are shown in Fig. 2.4. The fabrics after EPD need to be stored carefully to avoid contamination and are ready to be fabricated into FRPs with resin after they are completely dried. Several fabrication methods are available for thermosetting resins, while the deposited density and nanoadditive distribution may be changed during the fabrication process. To date, extensive researches have been conducted to deposit synthetic and organic nanoadditives onto fabrics by EPD; researches on natural inorganic nanoadditives are limited to the best of our knowledge.

This chapter deals with the EPD process for fabricating high-performance FRPs with hierarchical structure and mainly focuses on the deposition mechanics and kinetics of EPD; synergy effects of using surfactant; surface modification on HNTs and fabrics. Also, the

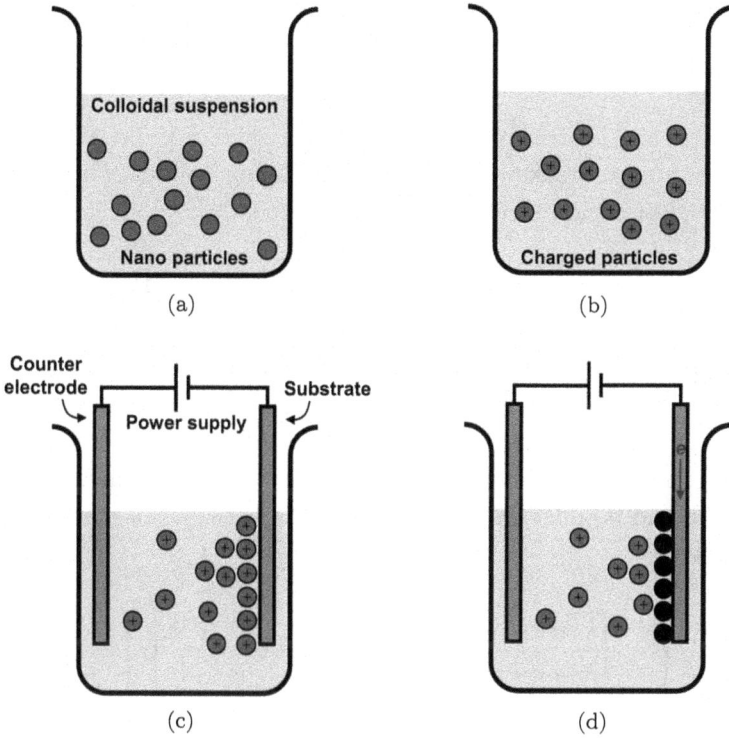

Fig. 2.4 Schematic of nanoadditives dispersion, charging, electrophoresis and deposition.[20]

inventive hybrid nanoadditive deposition and layer-by-layer assembly technique derived from the conventional EPD technique are briefly introduced.

2.2 Mechanical Approach for Hierarchical Structure

The mesoscale mechanical models for aligned and randomly distributed nanocomposites have been researched to predict the stiffness properties since the ellipsoid inclusion problems were proposed by Eshelby in 1957.[11] The Mori–Tanaka model, as the constitutive equations of the nanocomposites, is extensively used to solve the ellipsoid problems.[21,22] Based on the Mori–Tanaka model, the average strain and stress can be expressed by

$$\langle \sigma \rangle = C \langle \varepsilon \rangle \tag{2.1}$$

The angle brackets infer the averaging value. The effective average elastic moduli C is given by

$$C = C^m + v_i(C^i - C^m)\tilde{A}[v_m I + v_i \tilde{A}] \tag{2.2}$$

where C^m and C^f are the fourth-order stiffness tensor of matrix and inclusions. v_m and v_i are the volume fraction of matrix and inclusions, respectively; I is the identity tensor. $\tilde{A}$ is the equivalent strain concentration factor. Conventionally, the polymer matrix is isotropic and homogeneous, and the stiffness tensor of matrix can be expressed by

$$
\begin{Bmatrix}
\sigma_{11} \\
\sigma_{22} \\
\sigma_{33} \\
\sigma_{23} \\
\sigma_{31} \\
\sigma_{12}
\end{Bmatrix}
=
\begin{bmatrix}
k+u & k-u & k-u & 0 & 0 & 0 \\
k-u & k+u & k-u & 0 & 0 & 0 \\
k-u & k-u & k+u & 0 & 0 & 0 \\
0 & 0 & 0 & 2u & 0 & 0 \\
0 & 0 & 0 & 0 & 2u & 0 \\
0 & 0 & 0 & 0 & 0 & 2u
\end{bmatrix}
\begin{Bmatrix}
\varepsilon_{11} \\
\varepsilon_{22} \\
\varepsilon_{33} \\
\varepsilon_{23} \\
\varepsilon_{31} \\
\varepsilon_{12}
\end{Bmatrix}
\tag{2.3}
$$

where k and u are the bulk modulus and shear modulus of matrix, respectively. The equivalent strain concentration factor can be determined by

$$\tilde{A} = [I + S(C^m)^{-1}(C^i - C^m)]^{-1} \tag{2.4}$$

where S is the fourth-order Eshelby tensor associating with the shape of inclusions. The stiffness tensor C can be determined by substituting Eq. (2.4) into Eq. (2.2). For the randomly distributed inclusions, the orientation averaging should be taken into account:[23]

$$\langle C_{ijkl} \rangle = \frac{1}{4\pi} \int_{-\pi}^{\pi} \int_{0}^{\pi} C_{ijkl}(\theta, \phi) \sin \phi d\phi d\theta \qquad (2.5)$$

$$C_{ijkl} = a_{ip} a_{jq} a_{kr} a_{ls} C_{pqrs} \qquad (2.6)$$

where ϕ and θ are the Euler angles; $\boldsymbol{a}_{ij}$ is the transformation matrix mapping the jth unprimed to the ith primed axis and is defined by

$$a_{ij} = \begin{bmatrix} mp & -n & mq \\ np & m & nq \\ -q & 0 & p \end{bmatrix} \qquad (2.7)$$

where $m = \cos\theta$, $n = \sin\theta$, $p = \cos\phi$ and $q = \sin\phi$. As the inclusion is distributed uniformly and randomly in the macroscale, the stiffness tensor $\langle \boldsymbol{C} \rangle$ should be diagonally symmetric, which can be achieved by orientation averaging. Note that in hierarchical structure, nanoadditives mainly distribute in the fiber–matrix interfacial region, the stiffness property of FRPs with hierarchical distributed nanoadditives can be obtained by an innovative multi-step Mori–Tanaka modeling method derived from existing researches: (i) Firstly, take the representative volume element (RVE) from the hierarchical-structured interfacial area, and conduct the orientation averaging of the stiffness tensor of RVE. A diagonal symmetry tensor $\langle \boldsymbol{C}^i \rangle$ can be obtained; (ii) The RVE becomes the interface region with the surrounding fiber, treat the isotropic interface region as the equivalent matrix, and the stiffness tensor can be defined by substituting $\langle \boldsymbol{C}^i \rangle$ into Equation (2.2) as $\boldsymbol{C}^m$.[24–27] The equivalent strain concentration factor $\tilde{\boldsymbol{A}}$ also needs to be recalculated by using the Eshelby tensor of cylinder inclusions. At this step, the stiffness tensor of RVE, $\boldsymbol{C}^{i+f}$, is not necessary to conduct the orientation average; (iii) Selecting RVE including fiber, interfacial and polymer matrix. The composite stiffness tensor $\boldsymbol{C}^c$ can be determined by substituting $\boldsymbol{C}^{i+f}$ into Equation (2.2) as $\boldsymbol{C}^m$. The schematic representation of multi-step modeling of stiffness property of hierarchical structured FRPs is presented in Fig. 2.5.

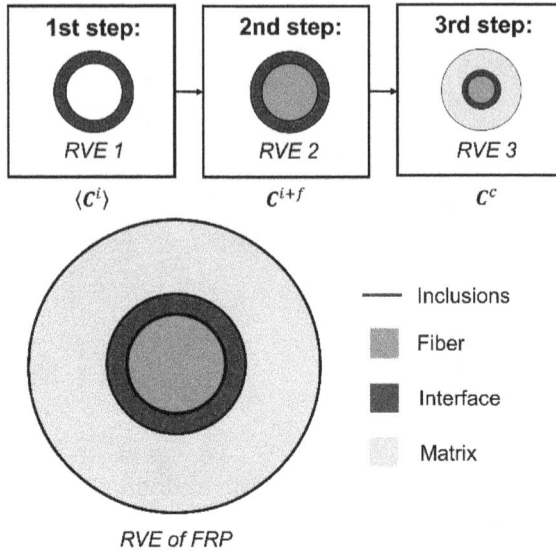

Fig. 2.5   Schematic representation of multi-step modeling of stiffness property of hierarchically structured FRP.

2.3   General Mechanisms of EPD

It is essential to understand how the nanoadditives migrate to the substrate having opposite charges. Firstly, once the electrical field is activated, the nanoadditives dispersed in suspension begin to move toward the substrate, and the electrical field acts as the "driving force." Then, the electrophoretic motion of the particles is stopped when they reach the substrate and lose their charges at the electrode surface. A closed loop of electronic forms with the generation of current. The discharged nanoadditives will still stay on the surface of the substrate and keep accumulating. The firmness of deposited nanoadditives depends on the field strength, zeta potential of electrophoretic bath and surface treatment of substrate fabrics, etc.; however, too large a field strength has the possibility to destabilize the nanoadditive movement, and eventually lead to the non-uniform deposition. The homogeneity of deposited nanoadditives strongly depends on the stability of the suspension. The stability of the suspension mainly refers to the electrostatic stability of the nanoadditives. On the one hand, the nanoadditives tend to be separated thanks to the repulsive electrostatic force; on the other hand, the attractive van der Waals (VDW) force tends to agglomerate the nanoadditives. The repulsive energy between two interacting nanoadditives can be determined based on the shape of the nanoadditives[18]:

$$E_R^{\text{Shpere}} = 2\pi a \int_x^{\infty} (G_{R,a}(x) - G_{R,a}(\infty)) \mathrm{d}x \qquad (2.8)$$

$$E_R^{\text{Flat}} = 2(G_{R,a}(x) - G_{R,a}(\infty)) \qquad (2.9)$$

where $G_{R,a}$ is the Gibbs free energy per unit of surface at distance x and distance ∞. The repulsive energy between two cylinder-shaped nanoadditives, which can be representative for HNTs, is more complicated:

$$E_R^{\text{Cylinder}} = 2 \int_x^{\infty} C_{\infty} RT (\cosh(\tilde{\psi}_m) - 1) \mathrm{d}x \qquad (2.10)$$

where $\tilde{\psi}_m$ is the dimensionless middle position potential; C_{∞} is the bulk ion concentration; R and T are the components of the equation of state.[28]

The attractive energy can be computed using Derjaguin *et al.* and Verwey *et al.*'s "DLVO" theory[29,30]:

$$E_A = -\frac{A}{6x} \frac{r_1 r_2}{r_1 + r_2} \qquad (2.11)$$

where A is the Hamaker constant, x is the separation of two nanoadditives; r_1 and r_2 are their radii. The total energy between the two identical nanoadditives in suspension can be eventually calculated as

$$E_T = E_A + E_R \qquad (2.12)$$

The plot of E_T is illustrated in Fig. 2.6. The attractive force dominates when the double-layer distance is close, and the attractive energy goes extremely large when two identical nanoadditives contact each other. Thus, the nanoadditives would be agglomerated and

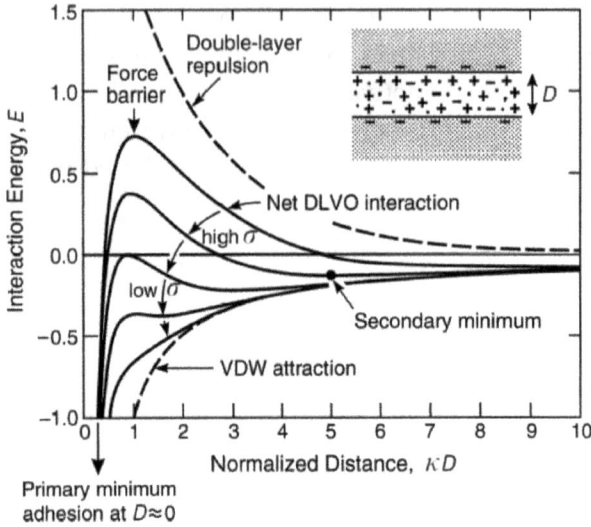

Fig. 2.6 Schematic DLVO plots including the repulsive electrostatic energy, VDW attraction energy and the total interaction energy.[31]

adhered irreversibly when they become too close. As the distance between the nanoadditives increases, the electrostatic repulsive force begins to dominate and form an energy peak, known as the "energy barrier." There may exist a secondary minimum of interaction after the energy barrier, but this peak is quite weak, and the agglomerate is reversible.

2.4 Kinetics of EPD

The deposited mass or thickness, which associates with parameters like, but not limited to, nanoadditive concentration, deposition time, the surface area of the substrate and the electric field, is essential for the quality of final product using EPD technique. Also, understanding the kinetics of EPD is helpful to control the desired deposit thickness and to determine the optimum EPD parameters. A kinetic model of EPD process proposed by is conventionally used.[32] Based on the mass conservation law, the deposited mass is equal to the mass removed from the suspension:

$$\frac{dw}{dt} = AuC \tag{2.13}$$

$$w = w_0 - VC \tag{2.14}$$

where C is the concentration of the nanoadditives; A is the area of the electrode; u is the velocity of the nanoadditives; t is the deposition time; w_0 is the initial weight of the nanoadditives inside the suspension; V is the suspension volume. On the base of electrophoretic theory, the velocity u of the solid particles is

$$u = \frac{\varepsilon\xi}{4\pi\eta}(E - \Delta E) \tag{2.15}$$

where ε is the dielectric constant of the liquid in suspension; η is the viscosity of the suspension; E is applied electric field; ΔE is the difference of E among the two electrodes; ξ is the zeta potential of nanoadditives in suspension. By substituting Equation (2.15) into Equation (2.13), the deposit rate can be expressed explicitly:

$$\frac{dw}{dt} = w_0 k e^{-kt} \tag{2.16}$$

where:

$$k = \frac{A}{V}\frac{\varepsilon\xi}{4\pi\eta}(E - \Delta E) \tag{2.17}$$

It is evident that the deposit rate $\frac{dw}{dt}$ is an exponential function of deposition time and the kinetic constant k. The deposit rate with different k values is illustrated in Fig. 2.8. According to Fig. 2.7, the increase of the kinetic constant is negligible for raising the maximum deposit rate, but can let the deposit rate reach the maximum value earlier. However, an increase of the kinetic constant also results in a faster decline of deposit rate after its maximum value. The total deposition mass, which can be obtained by integrating the deposit rate with regard to deposit time, is decreased with a large kinetic constant.

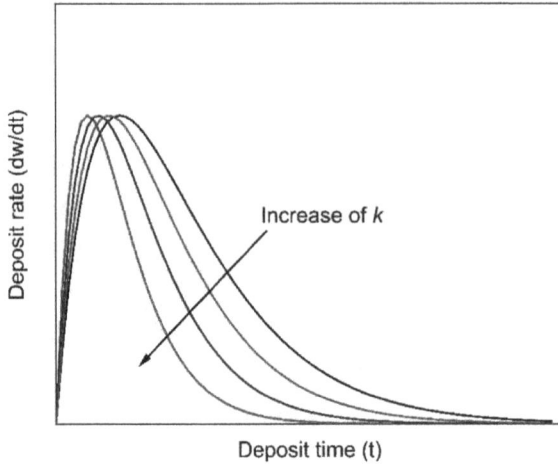

Fig. 2.7 Deposit rate with different kinetic constant k.

2.5 Fabrics Using in EPD

The fabrics always act as the substrate having few contributions to the EPD process, but there are still some modifiable items and attention points. Carbon fiber, which is conductive, is suitable for EPD as one electrode. However, carbon fabric cannot afford large current density because of the existence of warp and yarn. Empirically, unidirectional carbon fabric is more vulnerable to be damaged by current than woven fabrics because of the gap between carbon fiber strands. Other fibers like glass, basalt and aramid are non-conductive; it is impossible to directly employ them as the electrode. The solution is to place the non-conductive fabrics near the anode for the deposition of HNTs. The apparatus set up for non-conductive fabrics illustrated in Fig. 2.8 is referenced from studies of Mahmood *et al.*[33] During the EPD process, the HNTs moving toward the anode will be blocked by fabrics, eventually sediments on the fabric surface. The fabrics can also be treated or modified prior to EPD processes, like removal of sizing, silane modifications, etc. The removal of sizing makes HNTs deposit directly onto the substrate fabrics rather than the commercial sizing on the fiber surface. However, once the sizing is removed, the fabric becomes

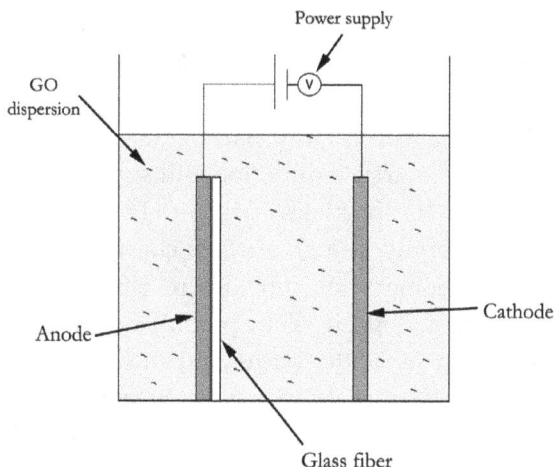

Fig. 2.8 Schematics of the EPD process of GO on GF.[33]

extremely difficult to handle and manipulate, and needs careful attention. Silane modifications can graft some desired organic groups onto the fabric, which usually aim to improve the interfacial adhesion, compatibility or other desired functions.

2.6 Deposition of HNTs as Reinforcing Additives

Formation of a stable suspension is essential for EPD process because the motionless and accumulated particles can result in the gradient in deposition. HNTs exist as tetrahedral siloxane on the external sheet and octahedral aluminols on the internal sheet, which has been described in Fig. 2.1. Low contents of hydroxyl groups on the surface of HNTs make them relatively hydrophobic as compared to other nanoadditives. The hydrophobic feature of HNTs makes them readily disperse in many non-polar polymers without extra modification steps. However, to homogeneously disperse HNTs in aqueous solutions is a challenge prior to EPD process. The dispersity of HNTs can be improved by adding some organic solvent, ethanol for example, into the aqueous solution. The interactions between HNTs are relatively weak compared to other nanoadditives like CNTs due to the limited -OH groups at the outer surface of HNTs. Also, the hollow tubular structure of HNTs reduces the large area intertube contact probability. The weak intertube interaction is favorable for the dispersity in both aqueous solutions and organic solvents.[34, 35] Zeta potential governs several key parameters in EPD, such as the density of the deposit, particle direction and speed, and the repulsive interactions between the particles, which determine the stability of the suspension. Generally, a high surface charge is needed not only to avoid particle agglomeration but also to enable the formation of a dense, highly packed deposit. Zeta potential manipulation is done through the addition of acids, bases and adsorbed ions.[20] Due to the outer surface consisting of layers with a SiO_2 bond, it will have a weak negative charge, whereas the inner lumen, which consists of layers with Al_2O_3, will have a strong positive charge.[36] At the appropriate pH range, the outer surface will be negatively charged while the inner lumen will be positively charged. The zeta potential of HNTs in a wide pH range presented in Fig. 2.9 is referenced from Viviana *et al.*[37] The zeta potential of HNTs is positive under pH = 2 and transfers to negative when pH is beyond 3. The HNTs become more negatively charged with larger pH values. Note that the strong negative charge of HNTs in neutral solution (pH = 7) provides HNTs

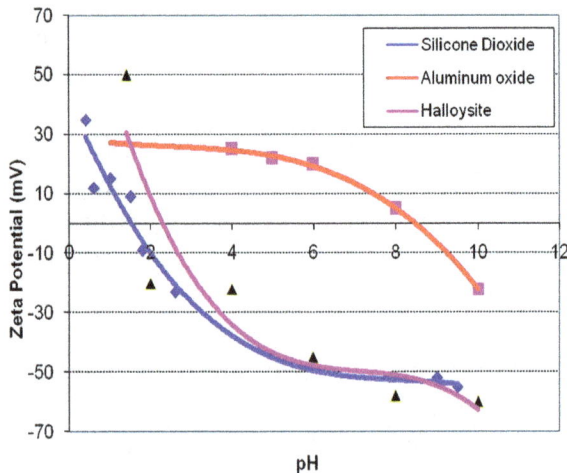

Fig. 2.9 Comparison of zeta potential curves for HNTs, silica and alumina nanotube.[37]

strong motivation under the electrical field even in pure aqueous solution. Also, the aqueous solution provides moderate dielectric constants and low viscosity, which ensure the migration effectiveness of HNTs. Once the HNTs are uniformly dispersed, the suspension is ready to perform as the electrophoretic bath due to their strong negative charge. Based on the prescribed properties of HNTs, it is reasonable to approve the feasibility of depositing HNTs by EPD.

Experimental work was conducted to investigate the mechanical properties of the CFRPs with the incorporation of HNTs by the EPD process.[38] The HNTs and SDS were both supplied by Sigma-Aldrich Co., Ltd. Carbon fiber woven plain (supplied by GM Composite Co., Ltd) was used as the reinforcement. About 0.2 wt% HNTs and 0.2 wt% SDS were first deposited with distilled water as the electrophoretic bath, followed by 5 h magnetic stirring and 20 min ultrasonic stirring for achieving homogeneous disparity. During the EPD process, the carbon fabrics were fixed in the acrylic frame and connected to the anode, where the cathode was connected to a piece of alloy gauze. The distance between two electrodes was a constant value of 10 mm. The voltage was set at a variable range from 0 V to 20 V, and the deposition time was set to a constant 5 min. After deposition, the carbon fabrics were dried on a hot plate at 200°C. CFRP samples, including 12 carbon fabric plies, were fabricated using the vacuum-assisted resin transfer molding (VaRTM) process. As a control group, the HNTs were mixed with plain resin with the same HNT loading and then infused through the as-received carbon fabrics. The curing process was conducted in the oven at 80°C for 12 h.

The bending strength and modulus of CFRPs with different experimental conditions are shown in Fig. 2.10. The ILSS of CFRPs is shown in Fig. 2.11. The bending strength of CFRPs exhibited less deviation, but the tendency could still be distinguished where the 15 V EPD-modified CFRPs performed better than any other with its ultimate strength of 670 MPa. On the contrary, the tangent modulus of elasticity separated an enormous range, and there was a clear tendency of the modulus of CFRPs increasing with the voltage intensity from 0 V to 15 V and decreasing at 20 V. For the short beam shear (SBS) test, the highest ILSS of 84 MPa value appeared

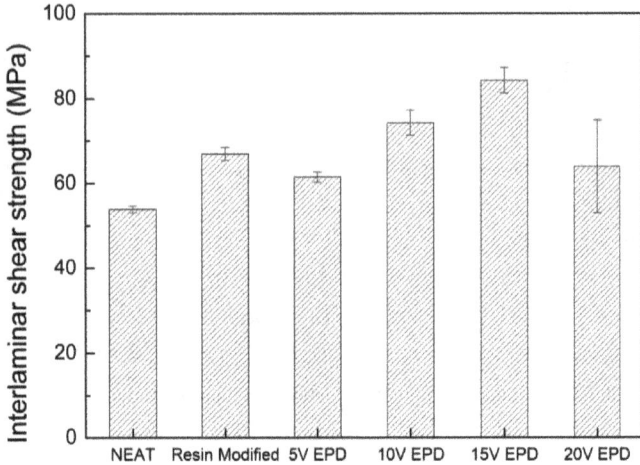

Fig. 2.10 ILSS of HNTs incorporated CFRPs with regard to different voltages in EPD.[38]

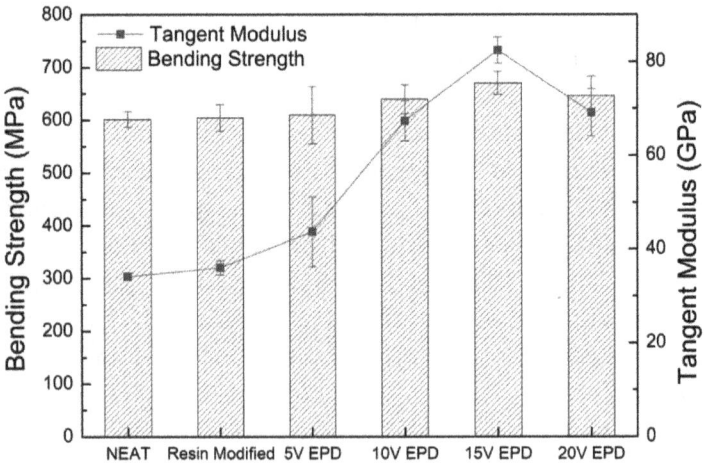

Fig. 2.11 Bending strength and modulus of HNTs incorporated CFRPs with regard to different voltages in EPD.[38]

at 15 V modified CFRPs, and the lowest value of 54 MPa appeared at neat CFRPs.

Mode I fracture toughness (G_{IC}) and Mode II fracture toughness (G_{IIC}) were obtained through double cantilever beam (DCB) and

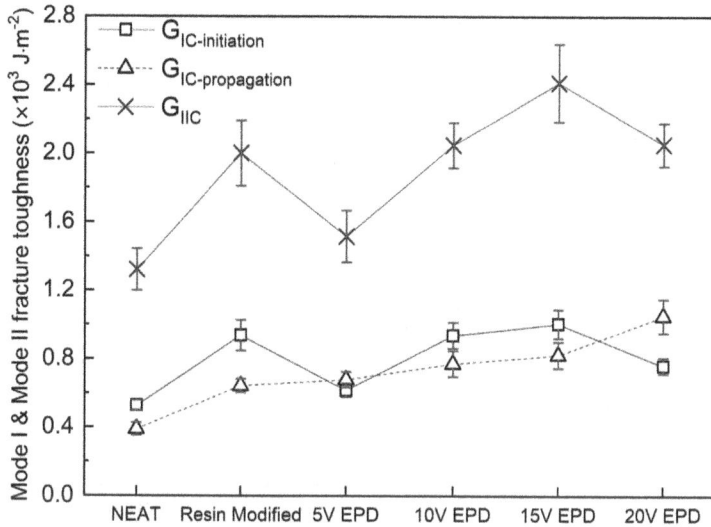

Fig. 2.12 Mode I and mode II fracture toughness of HNTs incorporated CFRPs with regard to different voltages in EPD.[38]

end-notched flexure (ENF) technique, respectively. The results of Modes I and II fracture toughness are illustrated in Fig. 2.12 and the representative load-displacement curves of DCB test are shown in Fig. 2.13. The fracture toughness at 15 V modified CFRPs was almost twice higher than that of neat CFRPs, which represented the feasibility and reliable performance on fracture toughness enhancement. The resin-modified specimens showed the second highest value caused by the excellent disparity of HNTs in epoxy. The GIC propagation showed a harsh tendency to rely on the voltage, which was thought of as significant as the amount of HNTs is more critical for resisting the secondary crack propagation rather than the homogeneity. The 15 V modified CFRPs showed both high value at crack propagation point and the smooth curve resulting in high GIC-initiation and GIC-propagation values. The neat CFRPs performed lower critical crack propagation load and multi-stage crack propagation resulting in the lowest fracture toughness.

The section morphologies of CFRPs with different modification conditions after SBS test photographed by SEM are shown

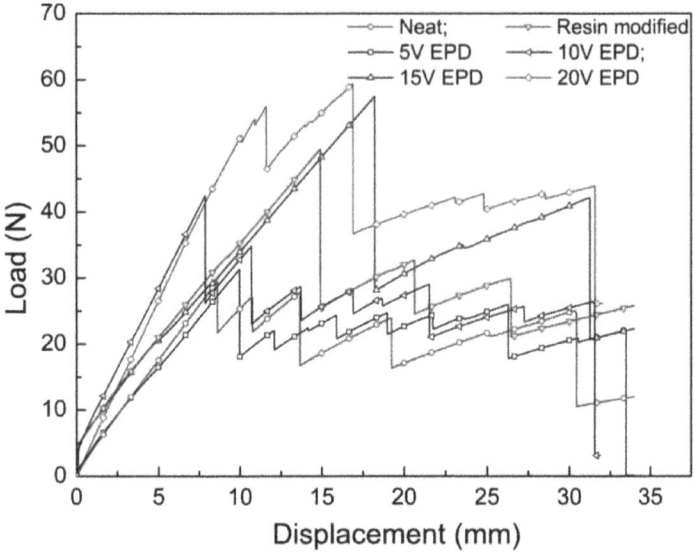

Fig. 2.13 Representative load-displacement curves of HNTs incorporated CFRPs with regard to different voltages in EPD.[38]

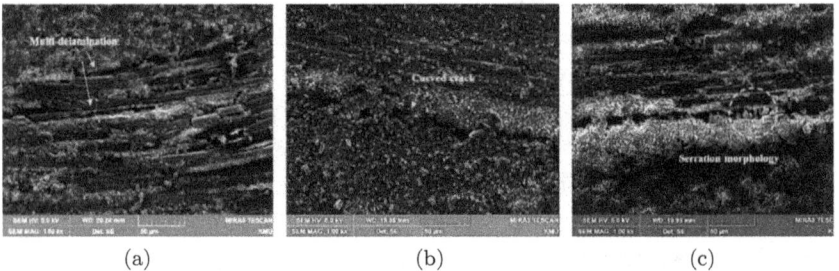

(a)   (b)   (c)

Fig. 2.14 SEM photographs of fractured ILSS specimen surfaces from: neat (a), resin modified (b) and 15 V modified (c) CFRPs.[38]

in Fig. 2.14. The neat CFRPs showed a multi-delamination phenomenon on the interface of fiber and matrix. Resin-modified CFRPs showed the curved crack path, which elongated the crack propagation length. The 15 V modified CFRPs presented the serration morphology. The high fracture toughness of 15 V modified CFRPs was recognized as the consequence of the dense adhesion of fiber and matrix that had excellent crack propagation resisting ability.

In conclusion, the 15 V modified CFRPs presented outstanding through-thickness properties. Furthermore, the through-thickness properties increased with the increase in voltage intensity from 0 V to 15 V and decreased at 20 V. The decrease was considered as the damage on carbon fabric caused by excessive voltage and heterogeneous HNTs dispersion. The resin-modified CFRPs also showed excellent properties because of the hydrophilic nature of HNTs. The SEM observation of 15 V modified CFRPs showed the serration morphology, which indicated the dense adhesion of fiber and matrix for impeding the crack propagation.

2.7 Particle Accumulation of EPD-Deposited HNTs During Fabrication

VaRTM is a well-established fabrication technique for FRPs. Nevertheless, using VaRTM to fabricate nanoparticles incorporated FRPs may lead to the inhomogeneous dispersion of nanoparticles in the direction of resin injection. This effect is caused by the anchoring effect of carbon fabrics, and by the scouring effect of resin flow. Specifically, the anchoring effect of carbon fabrics can hold the nanoparticles in their original positions, thus obstructing the nanoparticles from flowing freely within the resin. The scouring effect of resin flow is the phenomenon whereby the resin flow's front "scours" the nanoparticles toward the outlet. It is critical to explicate these two effects for further applications in nanocomposite manufacturing. The anchoring and scouring effects during the fabrication process of amorphous HNT (A-HNT), which was successfully synthesized and reported in Kim et al.[39] and Park et al.[40] were introduced in this chapter from the aspect of the mechanical properties.[41]

The HNTs were synthesized to A-HNT by annealing the neat HNTs at 1000°C for 4 h in an electric furnace. When produced by oxidative dehydrogenation, A-HNTs form amorphous silicon dioxide (SiO_2) and gamma aluminum oxide (γ-Al_2O_3). In the resin-blending process, epoxy resin was blended with 0, 0.2, 0.5, 0.8 and 1.0 wt% A-HNTs using an ultrasonic homogenizer, prior to infusion. During the EPD process, 0.2, 0.5, 0.8 and 1.0 wt% A-HNTs with identical

amounts of sodium dodecyl sulfate (SDS) were dispersed in distilled
water. They were then deposited onto carbon fabrics under 750 V/m
field strength, during a 5 min period. CFRPs with 12 laminæ
were fabricated through VaRTM on a flat alloy mold. The resin-
blended and EPD-modified experimental groups are labeled herein
as "resin blended group (RG)" and "EPD-modified group (EG),"
respectively, prior to their specific A-HNT concentrations. The curing
process was conducted in an oven at 80°C for 4 h. The laminates
were evenly divided into three sections, vertical to the infusion
direction. To achieve an effective demonstration, these sections are
abbreviated herein as "sA," "sB" and "sC." Moreover, an additional
moniker of "R" or "E" is used herein to indicate the incorporation
method, when necessary. Schematics of the VaRTM assembly and
the section division processes are shown in Fig. 2.15. Bending, SBS
and ENF tests were implemented to determine the through-thickness
properties of each section.

The bending, ILSS and mode II fracture toughness results are
illustrated in Figs. 2.16–2.18, respectively. The highest bending
strength occurred when 0.5 wt% A-HNTs were incorporated, for
both RG and EG. Both incorporation methods showed very similar
tendencies, in that their bending strengths increased before 0.5 wt%,
then decreased until 1.0 wt%. The relatively lower bending strength

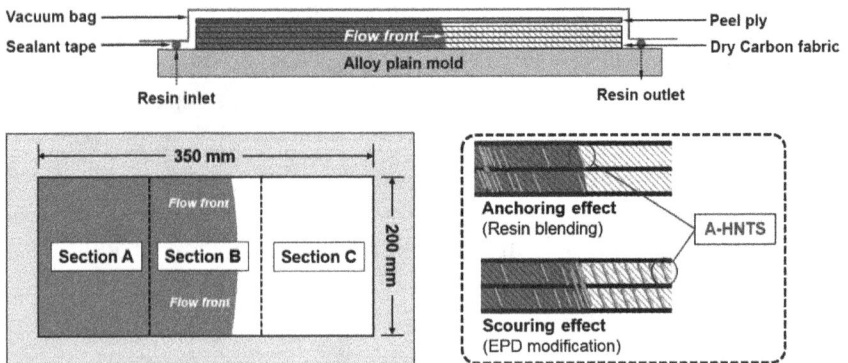

Fig. 2.15 Schematic representation of anchoring and scouring effects during
VaRTM.[41]

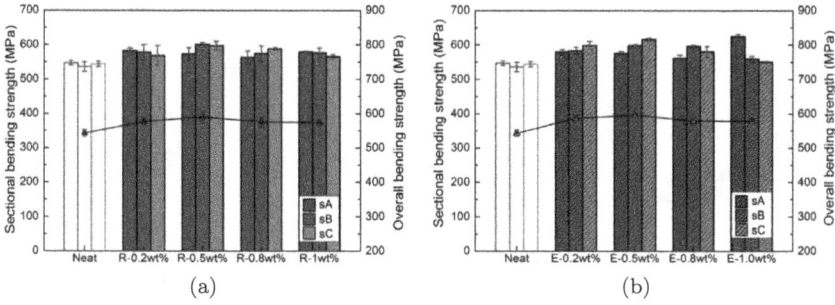

Fig. 2.16 Bending strength of CFRP laminates with different A-HNT contents incorporated by resin-blending (a) and EPD (b).

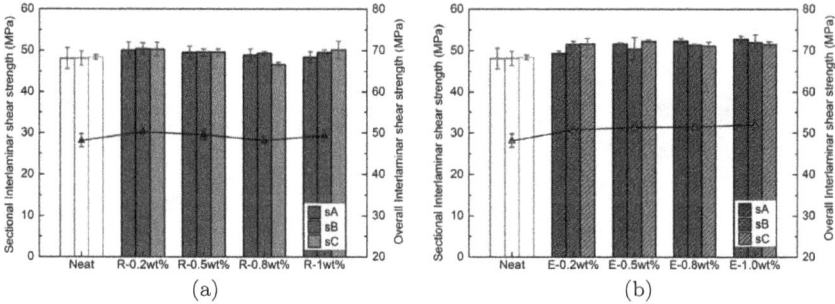

Fig. 2.17 ILSS of CFRP laminates with different A-HNT contents incorporated by resin-blending (a) and EPD (b).

enhancement observed in EG is in contrast to the results of the previous study.[38] The reduced structural integrity of the A-HNTs leads to the observed decreases in the polarity and dispersion in the aqueous solution. This resulted in the relevantly low observed motivation and uniform deposition during the EPD process. The reversed "U" shape visible in the section strength demonstrates that there were shifts or offsets to the overall trends of the bending strengths. The bending strength of sAR shifted to the left (i.e., a lower concentration of A-HNTs), and the bending strength of sCR shifted to the right (higher concentration of A-HNTs).

Dense carbon filaments anchored the dispersed A-HNTs in the epoxy resin. The anchored A-HNTs tended to aggregate and form

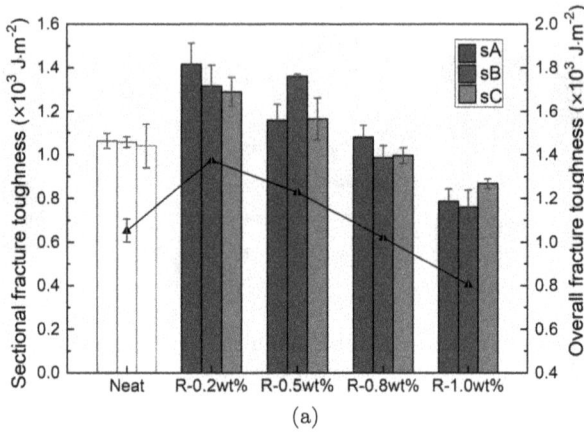

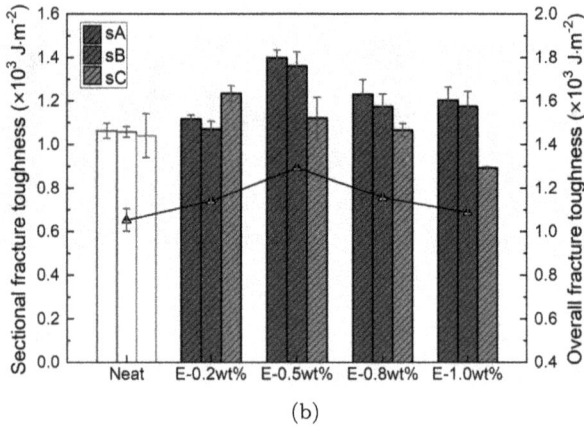

Fig. 2.18 G_{IIC} of CFRP laminates with different A-HNT contents incorporated by resin-blending (a) and EPD (b).

mesoscale clusters, which exacerbated the anchoring effect and generated stress concentration sources. As a consequence, the A-HNTs displayed a gradient decline distribution along the direction of the resin flow. The sAE exhibited a continuous increase until a concentration of 1.0 wt%, whereas the sCE exhibited an unreversed decline in bending strength. The deposited A-HNTs yielded the observed inclines in gradient distribution aligned with the resin infusion direction, which was caused by the resin scouring effect. The scouring effect can result in the extensive aggregation of A-HNTs at the end part of the laminate. This can change the resin fluidity and can form stress concentration sources. The ILSS values of RG and EG showed similar tendencies with regard to their bending strengths. The overall ILSS of EG showed a slight, but uninterrupted, increase up until E-1.0 wt%, whereas the ILSS of RG consisted of an unstable increase.

The G_{IIC} values show more spectacular changes across the different A-HNT concentrations and incorporation methods. The overall G_{IIC} reached its maximum value for R-0.2 wt%. It then decreased dramatically in a linear fashion, up until R-1.0 wt%. In Fig. 2.18(b), the overall G_{IIC} increased before E-0.5 wt%, before decreasing slightly. The lowest value, which also appeared at E-1.0 wt%, still exhibited an enhancement, compared to the neat CFRPs. Furthermore, the G_{IIC} exhibited higher sensitivity following the incorporation of A-HNTs; even a small amount of A-HNTs was shown to improve the G_{IIC} significantly, but extensive incorporation caused severe property deterioration. The A-HNTs were scoured and aggregated by the resin flow, which resulted in the observed declined gradient of G_{IIC} from sAE to sCE when extensive A-HNTs were incorporated.

The validation of the anchoring and scouring effects are significant for further applications of VaRTM in the fabrication process. However, the quantitative analysis of nanoparticle distribution during VaRTM still needs further studies. The surface morphology of fractured ENF tests are shown in Fig. 2.19. The incorporated A-HNTs of sAR-0.5 wt% are found to be dispersed in the bulk matrix

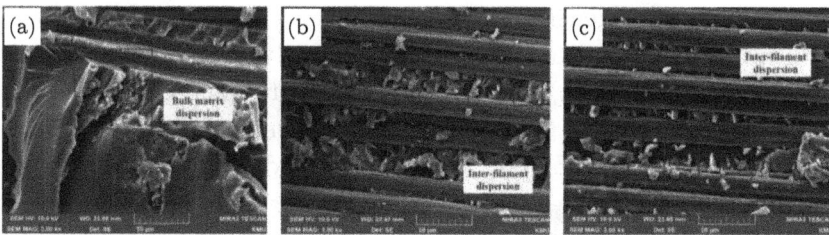

Fig. 2.19 Morphology of fractured sAR-0.5 wt.% (a), sAE-0.5 wt.% (b) and sCE-0.5 wt.% (c) ENF specimens.[39]

and aggregated. The A-HNT clusters in the matrix can impede and bend the crack propagation. However, the aggregated clusters can also become a source for stress concentration, and may, therefore, facilitate crack initiation. Moreover, the A-HNTs dispersed in the matrix provided only a negligible contribution to the bridging effect between the fiber and the matrix. The A-HNTs were dispersed in the interfilament areas in sAE-0.5 wt% and sCE-0.5 wt%, as shown in Fig. 2.19(b) and (c), respectively. Severer A-HNT accumulation can be observed in Fig. 2.19(c) due to the scouring effect. The unique interfilament dispersion occurred as a consequence of the EPD and resin scouring. The A-HNTs among sAE-0.5 wt% bridged the carbon fiber and matrix; thus, more of the residual matrix remained bonded to the fiber surface. This bridging between the fiber and the matrix effectively enhanced stress transfer, thus reducing delamination.

In conclusion, the overall bending strength, ILSS and G_{IIC} all showed a gradual decrease, after an initial ascent, in response to the increasing amounts of A-HNTs. During VaRTM, the A-HNTs in RG tended to accumulate near the resin inlet due to the anchoring effect caused by the dense carbon filaments. The A-HNTs in EG, however, tended to accumulate near the resin outlet, due to being influenced by resin scouring. The preserved properties observed in EG, following extensive incorporation, are attributable to the unique interfilament dispersion of A-HNTs.

2.8   Effects of Surfactants and Hybrid Deposition

Surfactants are generally organic amphiphilic molecules with hydrophilic and hydrophobic groups, usually amphiphilic organic compounds, containing hydrophobic and hydrophilic groups. Therefore, they are soluble in both organic solvents and aqueous solutions. Typically, a surfactant includes one hydrophilic head and one hydrophobic tail; the head is attracted to water, and the tail is attracted to oil. Usually, surfactants can be divided into different categories according to the polar head group: anionic, cationic, zwitterionic and non-ionic.[42]

During EPD process, surfactants have abilities to improve the suspension stability and increase the zeta-potential of suspension. As prescribed, individual HNT contains the negative-charged outer surface and positive-charged inner surface, the anionic and cationic surfactants are both applicable for HNTs, while the effect may exist with enormous deviation. Giuseppe *et al.* investigated the effect of both anionic and cationic surfactants on the colloid stability and solubilization ability of HNTs.[43] The authors chose sodium dodecanoate (NaL) as the cationic surfactant and decyltrimethy-lammonium bromide (DeTAB) as the anionic surfactant to perform the investigation. The DeTAB molecular was absorbed into the negative-charged outer surface of HNT and the NaL molecular was absorbed into the positive lumen. The schematic representation of the different absorption mechanisms is presented in Fig. 2.20. The results showed that the DeTAB/HNTs system tends to form aggregates due to the hydrophobic attractive interactions generated by the cationic surfactant adsorbed into the HNTs outer surface while this phenomenon is absent in the presence of NaL that adsorbs at the HNTs lumen.

Another research conducted by Lun *et al.* shows that adding SDS, as a kind of anionic surfactant can significantly improve the suspension stability by investigating the particle diameter distribution.[44] The authors found that the particle diameter distributed more uniformly after adding the SDS, which indicates the larger amount of dispersed HNTs. The particle diameter distribution conjoining with sedimentation observation is shown in Fig. 2.21. Also, the

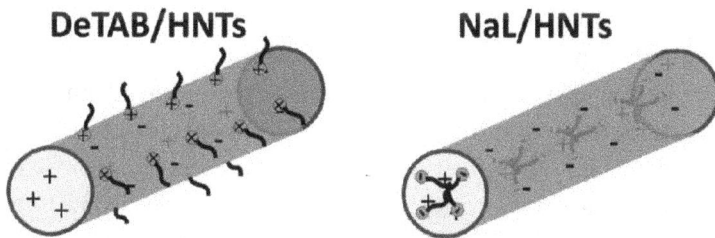

Fig. 2.20 Schematic reprsentation of different absorption mechanisms with HNTs.[43]

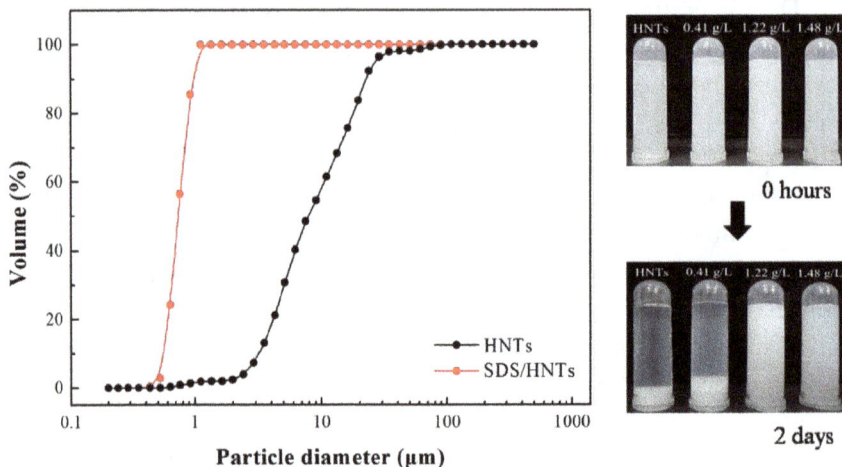

Fig. 2.21 HNTs particle diameter distribution conjoining with sedimentation observation.[44]

zeta potential of SDS/HNTs suspension was significantly increased. However, the research of Lun *et al.* and Cavallaro *et al.* derived different results on how the anionic surfactant influences the stability and zeta potential of HNTs-water suspension, and the synergy effects of surfactant with HNTs need further investigations.[43,44] Yu *et al.* opined that the anionic SDS molecules were first absorbed into the positive-charged inner lumen surface of HNTs by attracting electric force, and attached onto the outer surface by the VDW force when the concentration of SDS was high enough.[38] The schematic representation of SDS absorption mechanism of HNT is presented in Fig. 2.22.

Instead of depositing single nanoadditives, multi nanoadditives can also be dispersed in suspensions and deposited by EPD technique. The features of HNTs like modifiable external and internal sheets, different electric charges of the outer surface and inner lumen, high absorptivity due to tubular structure and biocompatibility make HNTs the ideal candidate for hybrid deposition. Bertolino *et al.* performed a hybrid deposition of three kinds of biopolymers (methylcellulose, alginate, chitosan) with HNTs and prepared the nanocomposites.[45] The authors found that the cationic chitosan

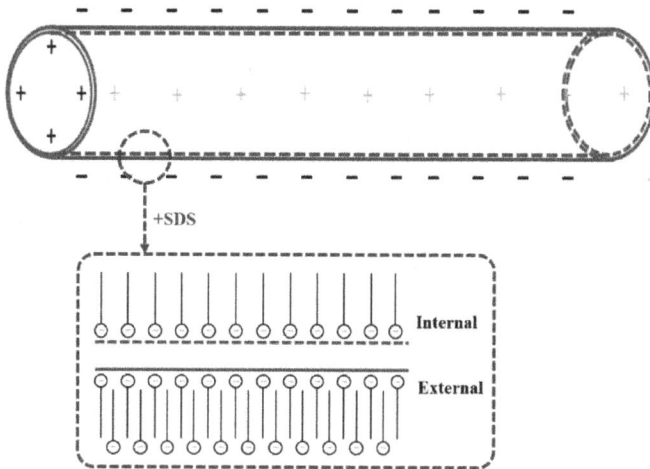

Fig. 2.22 Schematic of SDS absorption mechanism of HNT.[38]

Fig. 2.23 Schematic representation of the interaction of HNTs with three different biopolymers.[45]

attached onto the outer surface of HNTs and the anionic alginate was absorbed into the inner lumen. The non-ionic methylcellulose shows no specific distribution on inner and outer surfaces. The schematic representation of the interaction of HNTs with three different biopolymers is presented in Fig. 2.23. The mechanical properties of nanocomposites are also strongly connected to the interactions between biopolymer/nanofiller. The alginate encapsulation into the HNT lumen improved the tensile properties of the nanocomposites. Deen *et al.* also prepared the chitosan–HNT film for anti-corrosion usage using EPD process.[46] The cationic chitosan acted as the dispersing and charging agent.

2.9 Conclusion

Combined with empirical, theoretical and mechanistic studies, this chapter introduced in detail the EPD technique for fabricating FRPs with hierarchically distributed HNTs. Hierarchical structures of FRPs have been drawing attention since the last decade and have been proved to be effective in enhancing the interfacial and interlaminar properties. However, due to the ambiguous understanding of the fiber–matrix interfacial, the mechanical model of hierarchical structure has not been established yet. HNTs have huge modification potentials due to the unique molecular structure, while the applications of HNTs on FRPs are still limited. Although EPD is a mature technique, the most significant task for implementing EPD in nanotechnology is to prepare the stable and highly charged suspension. Both sweat and inspiration are indispensable to achieve further understanding of any scientific discipline.

References

[1] G. Dehm, B. N. Jaya, R. Raghavan and C. Kirchlechner, *Acta Mater.*, 142, 2018, 248–282.
[2] S. Prashanth, K. Subbaya, K. Nithin and S. Sachhidananda, *J. Mater. Sci. Eng.*, 6, 2017, 2–6.
[3] M. A. Karataş and H. Gökkaya, *Def. Technol.*, 14, 2018, 318–326.
[4] J. F. Rakow and A. M. Pettinger, Failure Analysis of Composite Structures in Aircraft Accidents, in ISASI Annual Air Safety Seminar Cancun, 2006, 11-4. ISASI.
[5] R. Das, B. F. Leo and F. Murphy, *Nanoscale. Res. Lett.*, 13, 2018, 183.
[6] R. Kamble, M. Ghag, S. Gaikawad and B. K. Panda, *J. Adv. Res.*, 3, 2012, 196.
[7] Y. Ye, H. Chen, J. Wu and C. M. Chan, *Compos. Sci. Technol.*, 71, 2011, 717–723.
[8] K. Prashantha, H. Schmitt, M.-F. Lacrampe and P. Krawczak, *Compos. Sci. Technol.*, 71, 2011, 1859–1866.
[9] Y. Dong, J. Marshall, H. J. Haroosh, S. Mohammadzadehmoghadam, D. Liu, X. Qi and K.-T. Lau, *Compos. Part A Appl. Sci. Manuf.*, 76, 2015, 28–36.
[10] Y. Joo, Y. Jeon, S. U. Lee, J. H. Sim, J. Ryu, S. Lee, H. Lee and D. Sohn, *J. Phys. Chem. C.*, 116, 2012, 18230–18235.
[11] J. D. Eshelby, *P. Roy. Soc. A Math. Phy.*, 241, 1957, 376–396.
[12] J. Karger-Kocsis, H. Mahmood and A. Pegoretti, *Compos. Sci. Technol.*, 186, 2020, 107932.

[13] M. Sharma, S. Gao, E. Mäder, H. Sharma, L. Y. Wei and J. Bijwe, *Compos. Sci. Technol.*, 102, 2014, 35–50.

[14] J. Karger-Kocsis, H. Mahmood and A. Pegoretti, *Prog. Mater. Sci.*, 73, 2015, 1–43.

[15] J. J. Ku-Herrera, O. F. Pacheco-Salazar, C. R. Ríos-Soberanis, G. Domínguez-Rodríguez and F. Avilés, *Sensors-Basel*, 16, 2016, 400.

[16] H. Qian, E. S. Greenhalgh, M. S. Shaffer and A. Bismarck, *J. Mater. Chem.*, 20, 2010, 4751–4762.

[17] P. Amrollahi, J. S. Krasinski, R. Vaidyanathan, L. Tayebi and D. Vashaee, *Electrophoretic Deposition (EPD): Fundamentals and Applications from Nano-to Microscale Structures*, Springer International Publishing, 2015.

[18] P. Sarkar and P. S. Nicholson, *J. Am. Ceram. Soc.*, 79, 1996, 1987–2002.

[19] L. Besra and M. Liu, *Prog. Mater. Sci.*, 52, 2007, 1–61.

[20] P. Amrollahi, J. S. Krasinski, R. Vaidyanathan, L. Tayebi and D. Vashaee, *Electrophoretic Deposition (EPD): Fundamentals and Applications from Nano- to Microscale Structures*, Springer International Publishing, 2016.

[21] T. Mori and K. Tanaka, *Acta Metall.*, 21, 1973, 571–574.

[22] P. R. Budarapu, X. Zhuang, T. Rabczuk and S. P. Bordas, *Multiscale Modeling of Material Failure: Theory and Computational Methods*, Elsevier, 2019.

[23] J. H. Huang, *J. Mater. Sci. Eng. A.*, 315, 2001, 11–20.

[24] C. L. Tucker III and E. Liang, *Compos. Sci. Technol.*, 59, 1999, 655–671.

[25] Y. Cheng, K. Zhang, B. Liang, H. Cheng, G. Hou, G. Xu and W. Jin, *Int. J. Mech. Sci.*, 161, 2019, 105014.

[26] L. Poh, C. Della, S. Ying, C. Goh and Y. Li, *AIP Adv.*, 5, 2015, 097153.

[27] G. Tandon and G. Weng, *Compos. Sci. Technol.*, 27, 1986, 111–132.

[28] J.-Y. Choi, H.-B. Dong, S.-J. Haam and S.-Y. Lee, *Bull. Korean Chem. Soc.*, 29, 2008, 1131–1136.

[29] B. Derjaguin and L. Landau, *Zh. Eksp. Teor. Fiz.*, 11, 1945, 802–821.

[30] E. J. W. Verwey, *J. Phys. Chem. Lett.*, 51, 1947, 631–636.

[31] F. L. Leite, C. C. Bueno, A. L. Da Róz, E. C. Ziemath and O. N. Oliveira, *Int. J. Mol. Sci.*, 13, 2012, 12773–12856.

[32] Z. Zhang, Y. Huang and Z. Jiang, *J. Am. Ceram. Soc.*, 77, 1994, 1946–1949.

[33] H. Mahmood, M. Tripathi, N. Pugno and A. Pegoretti, *Compos. Sci. Technol.*, 126, 2016, 149–157.

[34] X.-L. Xie, Y.-W. Mai and X.-P. Zhou, *Mat. Sci. Eng. R.*, 49, 2005, 89–112.

[35] M. Du, B. Guo and D. Jia, *Polym. Int.*, 59, 2010, 574–582.

[36] K. A. Zahidah, S. Kakooei, M. C. Ismail and P. B. Raja, *Prog. Org. Coat.*, 111, 2017, 175–185.

[37] V. Vergaro, E. Abdullayev, Y. M. Lvov, A. Zeitoun, R. Cingolani, R. Rinaldi and S. Leporatti, *Biomacromolecules*, 11, 2010, 820–826.

[38] T. Yu, Z. Chen, S.-J. Park and Y.-H. Kim, *Mod. Phys. Lett. B.*, 33, 2019, 1940023.

[39] Y. Kim, S. Park, J. Choi, K. Moon and C. Bae, *Int. J. Mod. Phy. B.*, 32, 2018, 1840070.

[40] S. Park, J. Choi, J. Lee, T. Yu, Z. Chen, Y. Kim and C. Bae, *Mod. Phys. Lett. B.*, 33, 2019, 1940020.
[41] T. Yu, Y.-H. Kim and S.-J. Park, *Mod. Phys. Lett. B.*, 34, 2020, 2040006.
[42] N. Dave and T. Joshi, *Int. J. Appl. Chem.*, 13, 2017, 663–672.
[43] G. Cavallaro, G. Lazzara and S. Milioto, *J. Phys. Chem. C.*, 116, 2012, 21932–21938.
[44] H. Lun, J. Ouyang and H. Yang, *Phys. Chem. Miner.*, 41, 2014, 281–288.
[45] V. Bertolino, G. Cavallaro, G. Lazzara, M. Merli, S. Milioto, F. Parisi and L. Sciascia, *Ind. Eng. Chem. Res.*, 55, 2016, 7373–7380.
[46] I. Deen, X. Pang and I. Zhitomirsky, *Colloids Surf. A Physicochem. Eng. Asp.*, 410, 2012, 38–44.

Chapter 3

Plasma-Treated Carbon Black Nanofiller for Improved Dispersion and Mechanical Properties in Electrospun Complex Nanofibers

Tae-Gyu Kim[*,‡], Jing Jing Zhang[†,§] and Xing Yan Tan[†,¶]

*Department of Nanomechatronics Engineering,
Pusan National University, Busan, Republic of Korea
†Department of Nano Fusion Technology,
Pusan National University, Busan, Republic of Korea
‡tgkim@pusan.ac.kr
§jalf1314521@gg.com
¶txy511@outlook.com

Carbon blacks (CBs) are nanoscale particles that can be used as fillers in composite materials. Due to their wide variety of unique properties,they have been applied in polymer composite materials as reinforcing fillers, conductive agents, electric and/or thermal conductive components among others.[1,2] Morphology of nanostructure has a major role in the characteristics of the nanoscale products. For example, CB/polymer composites prepared in thin films can function as sensors,[3] electrodes[4] and electromagnetic interference shielding.[5] While CB/polymer composites prepared in nanofibers network structure provide flexibility and strength.[6] Regardless of the selection of the preparation methods, CBs in the polymer matrix must be well-dispersed for a successful blending of CB/polymer composites. Therefore, the study of interfacial interaction between CBs and polymer matrix is important to improve the dispersibility of CBs. For this purpose, various studies of surface treatments have been carried out, such as surfactant surface adsorption,[7] encapsulation,[8] surface chemical grafting,[9] surface oxidation,[10] plasma treatment, etc. In this chapter, plasma treatment will be the focus of this topic as it has the advantages of shorter reaction, environment-friendly processing and providing a wide range of different functional groups[11] over the other surface treatment as mentioned previously. Then, the plasma-treated CBs are used to fabricate CB/polymer composites by electrospinning, and the changes in mechanical strength with the dispersibility of CBs are further explored.

3.1 Introduction

3.1.1 *Plasma Treatment*

Plasma is one of the four fundamental states of matter, and was first described by chemist Irving Langmuir in the 1920s.[12] When a solid is heated sufficiently the thermal motion of the atoms breaks the crystal lattice structure apart, and usually, a liquid is formed. When a liquid is heated enough that the atoms vaporize off the surface faster than they can recondense, a gas is formed. When a gas is heated enough that the atoms collide and knock their electrons off in the process, a plasma is formed: the so-called "fourth state of matter." Unlike the other three states (solid, liquid and gas), plasma does not exist freely on the Earth's surface under normal conditions. Plasma is a cloud of protons, neutrons and electrons where all the electrons have come loose from their respective molecules and atoms, giving the plasma the ability to act as a whole rather than as a bunch of atoms. First and foremost, plasma is more like a gas than any other state of matter because the atoms are not in constant contact with each other, but it behaves differently from a gas. It has what scientists call collective behavior. This means that the plasma can flow like a liquid, or it can contain areas that are like clumps of atoms sticking together. It can only be artificially generated by heating or subjecting a neutral gas to a strong electromagnetic field to a point where an ionized gaseous substance becomes increasingly electrically conductive and long-range electromagnetic fields dominate the behavior of the matter.[13] Plasma processing of materials has been developed fast in recent years for its advantages such as it only affects the chemical and physical properties of the substrate's outmost layer without altering the bulk properties. The science of plasma physics was developed to provide an understanding of these naturally occurring plasmas and in furtherance of the quest for controlled nuclear fusion. Plasma science has now been used in a number of other practical applications, such as the etching of advanced semiconductor chips and the development of compact X-ray lasers (Fig. 3.1). Many of the conceptual tools developed during fundamental research on the plasma state, such as the theory of Hamiltonian chaos, have found wide application outside the plasma field. The application of plasma technology can greatly reduce requirements of water, steam, energy and chemicals. At the same time, less toxic substances or pollutants will be released during processing.

In scientific studies, the plasma was generated by physical vapor deposition (PVD-as shown in Fig. 3.2) or chemical vapor deposition (CVD-as shown in Fig. 3.3). The gas sources can be selected following the experimental objective. The working principle of PVD or CVD uses electric discharge plasmas in which the energy to sustain the ionized state is supplied by an externally applied electric field. Most of the applications involve the use of low-pressure plasmas. The energetic species in these plasmas are the free electrons. They gain energy from the electric field faster than the ions do and are thermally isolated from the atoms and molecules as far as elastic collisions

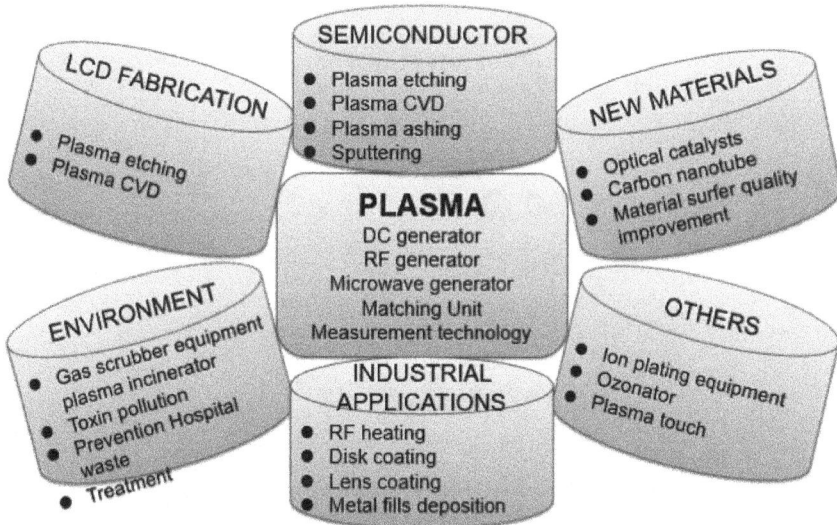

Fig. 3.1 Application fields of plasma treatment.

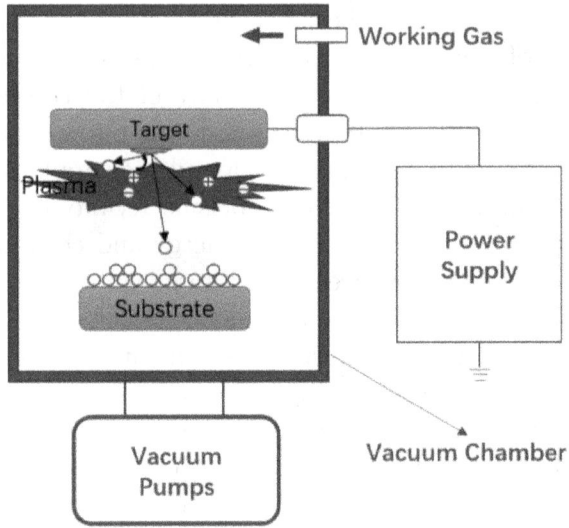

Fig. 3.2 Schematic representation of PVD.

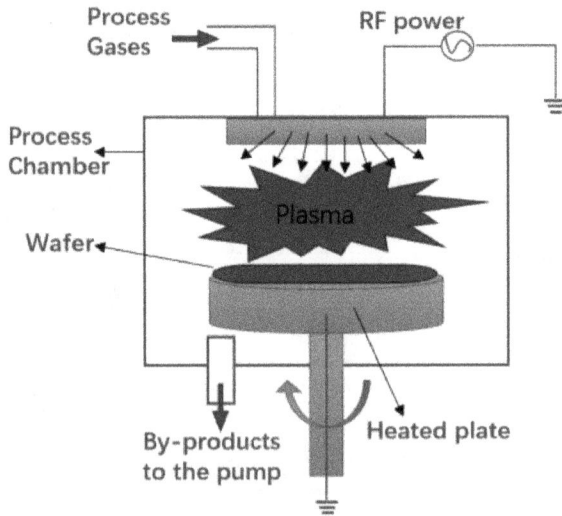

Fig. 3.3 Schematic representation of CVD.

are concerned, by the mass difference. Consequently, the electrons accumulate sufficient kinetic energy to undergo inelastic collisions and sustain the ionization, while the heavy particle (molecule) temperature remains low.

In recent years, plasma treatment is also used for improving the dispersibility of nanoscale powder as it can increase polar groups, including C–OH, COOH, C=O, O–C–O on the surface of nanoscale powder to increase the hydrophilic property and reduce the power of attraction between two particles.

3.1.2 *Electrospinning Process*

The electrospinning process is, in fact, a kind of fiber production method that uses electric force to draw charged threads of polymer solutions or polymer melts up to fiber diameters in the order of some hundred nanometers. The diameter of fibers fabricated by the electrospinning process can range from microns to nanometers. Its roots go back to the early 1930s when the process was patented by a researcher called Formhals. Electrospinning method characterizes both the electrospraying and the conventional solution dry spinning of fibers.[14] This kind of process does not require coagulation chemistry or high temperatures to produce solid threads from the solution. It makes the process particularly suited to the production of fibers using large and complex molecules. The advantage of the electrospinning process is its technical simplicity and its easy adaptability. It is based mainly on applying an electrical field by using a high voltage source between the tip of a nozzle and a collector in order to generate sufficient electrostatic force to overcome the surface tension in a droplet of polymer solution at the nozzle tip. Electrospinning from molten precursors is also practiced and is easy to operate; this method ensures that no solvent can be carried over into the final product.

The electrospinning process can be used in many fields. Its detailed applications are shown in Fig. 3.4. A set of simple electrospinning devices is shown in Fig. 3.5. The DC high voltage, needle, collector, etc., are the major components of this setup.

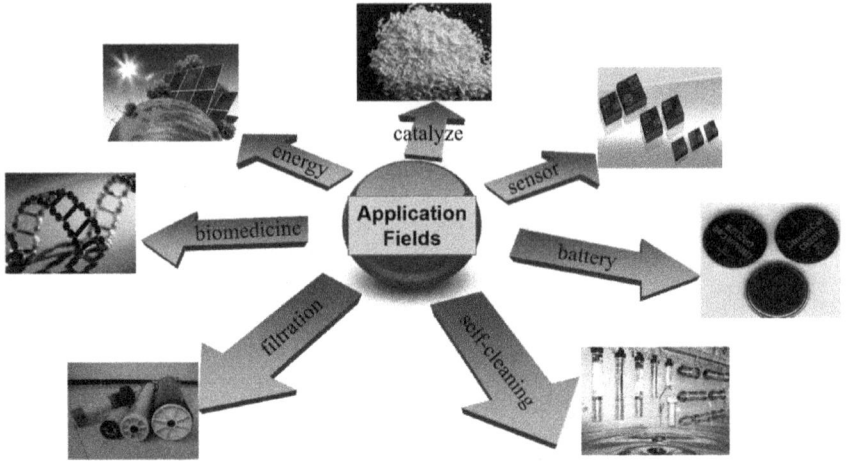

Fig. 3.4 Summary figure of electrospinning application fields.

Figure 3.6 shows the fiber formation by electrospinning and Taylor Cone, which is also mentioned in Fig. 3.5. The "Taylor Cone" is a key concept of the electrospinning process that refers to the cone observed in electrospinning. A jet of charged particles emanates above a threshold voltage. Aside from electrospray ionization in mass spectrometry, the Taylor Cone is very important in field-emission electric propulsion (FEEP) and colloid thrusters used in fine control and high efficiency (low power) thrust of spacecraft. The applied voltage determines the shape of the Taylor Cone. The suitable voltage is the voltage which can maintain the shape of Taylor Cone and keep electrospinning liquid jet steadily out. Figure 3.7 describes how the distribution of charge in the fiber changes as the fiber dries during a flight. As shown in Fig. 3.7, when sufficiently high voltage is applied to a liquid droplet, the body of the liquid becomes charged, the electrostatic repulsion counteracts the surface tension, and the droplet can be stretched. At the critical point, a jet of electrospinning liquid erupts from the surface. This point of eruption is what was mentioned before as Taylor Cone. If the molecular cohesion of the liquid is sufficiently high, stream breakup does not occur (if it does, droplets are electro sprayed), and a charged liquid jet is formed. As

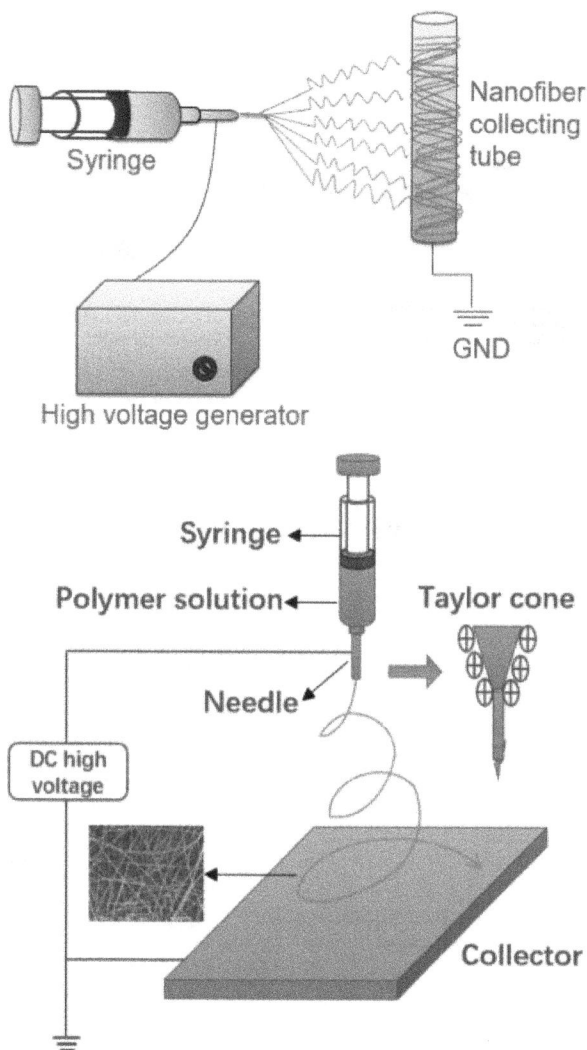

Fig. 3.5 Schematic demonstration of the electrospinning setup.

the jet dries during flight, the mode of current flow changes from ohmic to convective as the charge migrates to the surface of the fiber. The jet is then elongated by a whipping process caused by electrostatic repulsion initiated at small bends in the fiber until it is finally deposited on the grounded collector. The elongation and

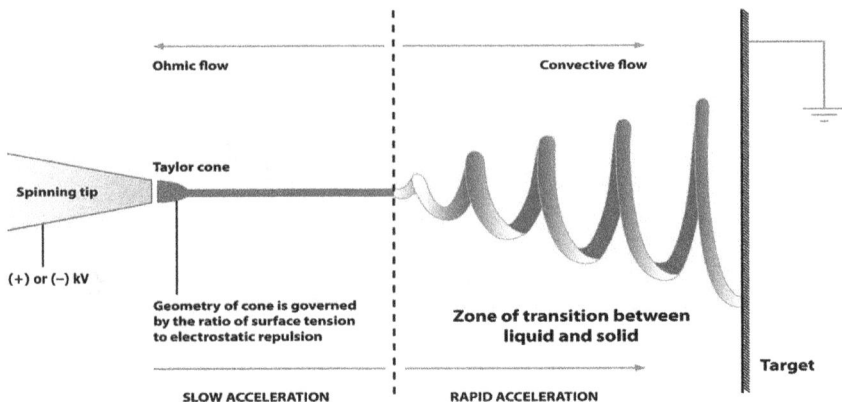

Fig. 3.6 The diagram showing fiber formation by electrospinning process.

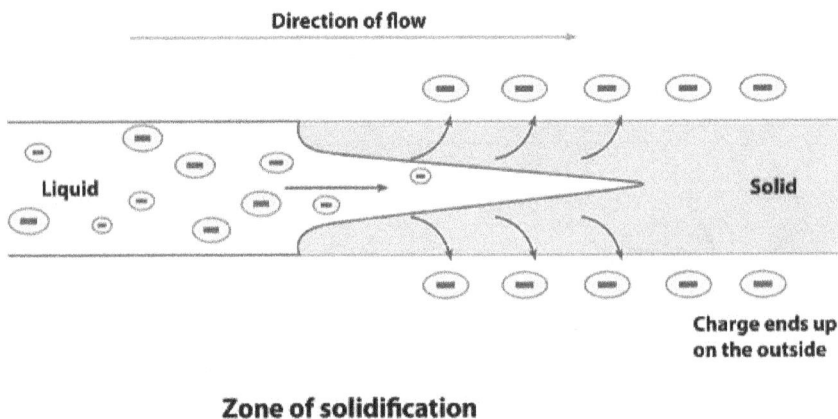

Fig. 3.7 Schematic representation showing how the distribution of charge in the fiber changes as the fiber dries during flight.

thinning of the fiber resulting from this bending instability lead to uniform fibers with nanometer-scale diameters.[15]

Electrospinning is a complicated process. There are several kinds of internal and external factors that can affect the morphology of the electrospinning product. During the preparation of a uniform form of fibers, the researchers need to consider multiple factors. A scholar called Taylor came up with a formula to study the influence

Table 3.1 Influence factors of the electrospinning process

Solution parameters	Experimental parameters	Ambient parameters
Polymer molecular weight	Electric potential	Temperature
Viscosity	Distance between the capillary and collection screen	Humidity
Concentration	Flow rate of the spinning solution	Air velocity in the chamber
Conductivity	Needle gauge	
Surface tension		

factors (see Equation (3.1)). Combining with this formula, we tried to list the principal influence factors of the electrospinning process in Table 3.1.

$$U_c^2 = 4\frac{H^2}{L^2}\left(\ln\frac{2L}{R} - \frac{3}{2}\right)(0.117\pi\gamma R) \qquad (3.1)$$

where U_c is the voltage, H is the distance from needle to the collector, L is the length of the needle, R is the radius of the needle and γ is the surface tension of the spinning solution.

3.2 Experiments

3.2.1 *Materials Used in the Research*

3.2.1.1 *Carbon Blacks (CBs)*

CBs are a form of amorphous carbon with a high surface-area-to-volume ratio, although lower than that of an activated carbon. The structural formula is shown in Fig. 3.8. The nature of the CBs' surface, porous structure, surface area and chemical composition are of vital importance.[16] It is dissimilar to soot in its much higher surface-area-to-volume ratio and significantly lower polycyclic aromatic hydrocarbon (PAH) content. However, CB is widely used as a model compound for diesel soot for diesel oxidation experiments. CBs

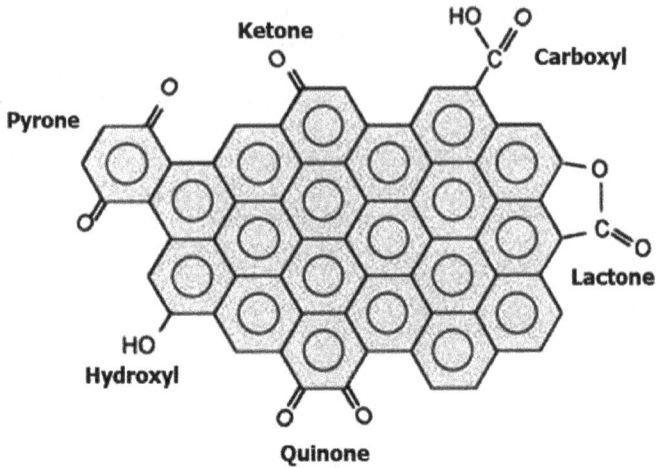

Fig. 3.8 The structural formula of CBs.

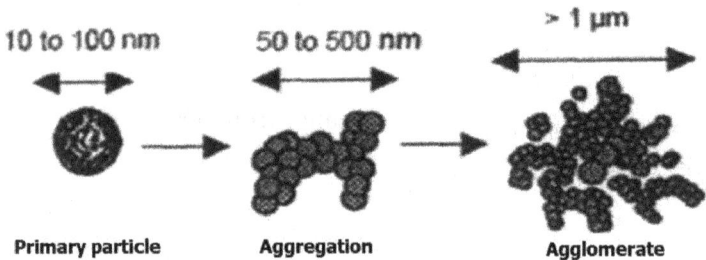

Fig. 3.9 The aggregation process of CBs.

are mainly used as a reinforcing filler in plastics, elastomers, paints and inks to modify the mechanical, optical and electrical properties of the materials. CB is also a kind of color pigment. Besides, CBs are electrically conductive and impart good conductivity to thermoplastic polymers.

The primary characteristics of CBs that influence the properties of CB compounds with elastomers are particle size, aggregate size, the morphology of the CB aggregates and their microstructure. The CB aggregates mentioned here are produced after small particles gather together. The aggregates further aggregate into agglomerates (as shown in Fig. 3.9). Because of aggregation, the applications of

CBs are limited. Therefore, the problem of dispersal presses for a solution.

The Chinese CB companies supplied the CBs used in this research. The primary particle size and specific surface area of the CBs are about 20 nm and 120 m^2/g, respectively.

3.2.1.2 *Poly (vinylidene fluoride-co-hexafluoropropene) (PVDF-HFP)*

PVDF-HFP is a copolymer that has been extensively studied as a gel polymer electrolyte.[17–19] It consists of crystalline poly (vinylidene difluoride) (PVDF), that can provide sufficient mechanical strength to form a free-standing film, and the amorphous hexafluoropropylene (HFP), that can absorb large amounts of liquid electrolyte to improve the ionic conductivity. Figure 3.10 shows the structural formula of PVDF-HFP. Due to the special structure, PVDF-HFP has great potentials in many applications such as battery electrode, waterproof membrane, etc.

In this experimental scheme, PVDF-HFP with molecular weight Mw = 400,000 g/mol (CAS No. 9011-17-0) was used as a solvent to prepare the electrospinning solution, and the dimethylformamide solution (DMF) (purity 99.8%, Daejung Chemicals and Metals Co., Korea) was used to dissolved PVDF-HFP.

3.2.1.3 *N,N-Dimethylformamide (DMF)*

DMF is a kind of organic compound whose formula is $(CH_3)_2NC(O)H$. The structural formula is shown in Fig. 3.11. It is a common solvent for chemical reactions and is a colorless, transparent

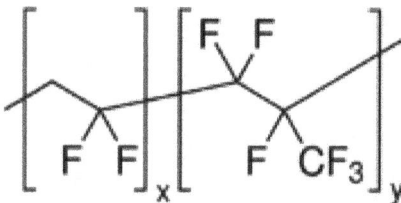

Fig. 3.10 The structural formula of PVDF-HFP.

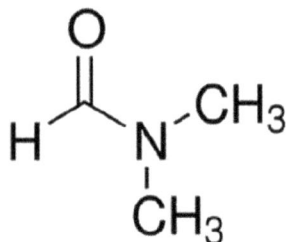

Fig. 3.11   The structural formula of DMF.

or weak yellow liquid. DMF is a polar (hydrophilic) aprotic solvent
with a high boiling point and has a special smell. It has toxic effects
on the skin, eyes and respiratory system. The primary use of the
DMF is as a solvent with a low evaporation rate. DMF is used
in the production of acrylic fibers and plastics. It also acts as a
solvent in peptide coupling for pharmaceuticals, in the manufacture
of adhesives, synthetic leathers, fibers, films and surface coatings,
and the development and production of pesticides.[17] Besides, DMF
has many uses in the research laboratory, such as a standard in
proton NMR spectroscopy allowing for a quantitative determination
of an unknown compound, a source of carbon monoxide ligands, the
solvothermal synthesis of Metal-Organic Frameworks and so on.

In this study, the DMF (purity 99.8%, Daejung Chemicals and
Metals Co., Korea) was used as a common electrospinning solvent to
dissolve PVDF-HFP polymer to prepare the electrospinning solution.

3.2.2   *Methods*

3.2.2.1   *Plasma Treatment of Nanoscale Powders*

To solve the dispersion problem of nanoscale powders and to improve
the interfacial interaction between CBs and polymer matrix, the
plasma treatment processing was introduced. The plasma treatment
processing for nanoscale CB powders was performed using the
current RF-PECVD equipment in the laboratory. The diameter and
height of the chamber are 600 mm and 800 mm, respectively, as shown
in Fig. 3.12(d). To reduce the dust emissions of CB powders in

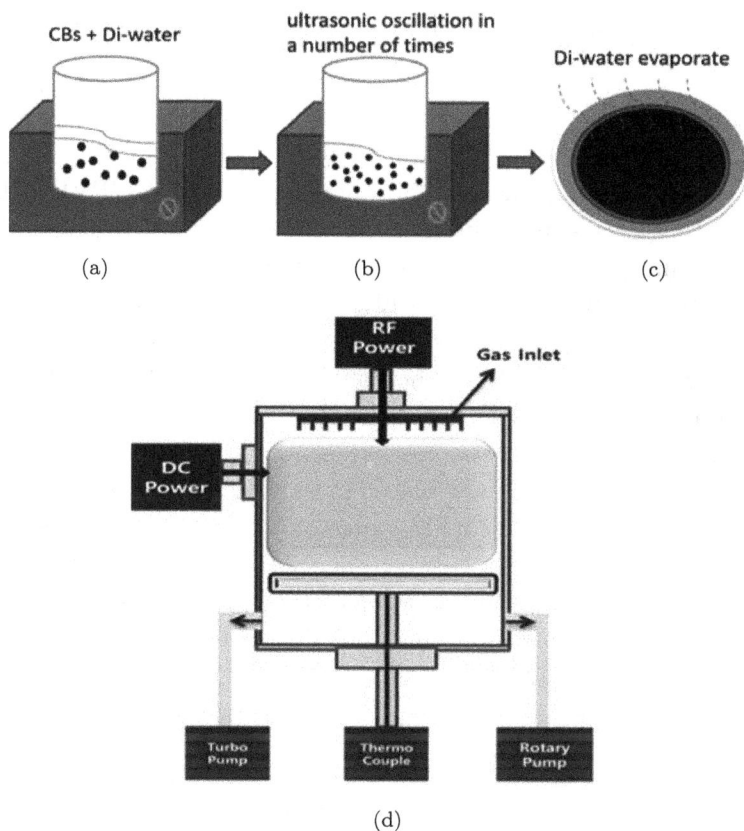

Fig. 3.12   Schematic diagram of plasma treatment.

the chamber, the powders needed to be pretreated into mud state
by using deionized water (Di-water). Figure 3.12(a)–(c) shows the
pretreatment process where the nanoscale powders were mixed with
Di-water by ultrasonic oscillation and after a period of oscillation
stirring the nanoscale powders turned into mud, then the CB mud
was spread evenly on the iron pan. Until the Di-water evaporated,
the iron pan was put into the chamber of RF-PECVD. The radio
frequency (13.56 MHz) was used capacitively. The chamber was first
evacuated to a pressure of about 10^{-4} Torr before plasma gas was
introduced. Then different kinds of gas plasma were used to treat the
muddy samples under the same conditions, respectively. The 300 W

of power was applied for 40 min, under the fixed pressure of 10^{-2} Torr and the flow rate of about 80 sccm.

3.2.2.2 *Electrospinning*

The electrospinning process produced all the CBs/PVDF-HFP composite systems used in the study. The solutions used in the electrospinning process were obtained firstly by a CBs/DMF suspension being held for 2 h in an ultrasonic bath. Then, PVDF-HFP polymer was dissolved in a stable suspension of CBs in the DMF utilization stirring method. The concentration of PVDF-HFP was maintained by 20 wt%. The mass fraction of CBs, CBs-N_2, CBs-O_2 was guaranteed as 1.7 wt%.

The polymer solution was placed in a syringe with a needle having an inner diameter of 1.26 mm (a gauge of N18). A voltage of ~17 kV was applied in a 16 cm gap between the spinneret and collection plate to obtain the randomly oriented electrospun nanofibers. A syringe pump fed the polymer solution at a flow rate range of about 2–3 ml/h. The overall lab equipment is shown in Fig. 3.13.

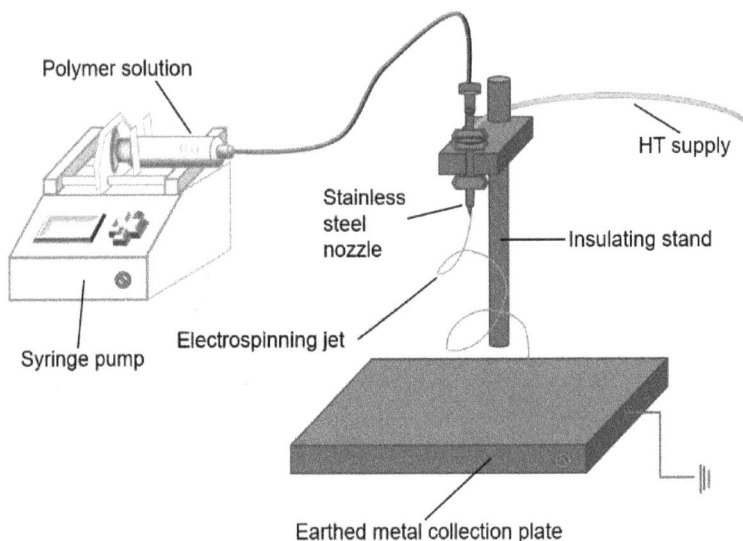

Fig. 3.13 Schematic diagram of the electrospinning process.

3.3 Results and Discussions

3.3.1 *Scanning Electron Microscope (SEM) of Nanoscale Powders*

SEM is a kind of electron microscope that can produce images of samples by scanning the surface with focused electrons. The main SEM components include the source of electrons, column down which electrons travel with electromagnetic lenses, an electron detector, sample chamber, computer and display to view the images. Electrons are produced at the top of the column, accelerated down and passed through a combination of lenses and apertures to produce a focused beam of electrons that hits the surface of the sample. The sample is mounted on a stage in the chamber area. Unless the microscope is designed to operate at low vacuums, both the column and the chamber are evacuated by a combination of pumps. The level of the vacuum will depend on the design of the microscope. The electrons interact with atoms in the samples, producing various signals that contain information about the samples' surface topography and com-position. The electron beam is scanned in a raster scan pattern, and the beam's position is combined with the detected signal to produce an image. SEM can achieve resolution better than 1 nanometer. To identify the feasibility of using plasma to improve dispersibility, the SEM images of nanoscale powders with and without N_2-plasma treatment dissolved in Di-water were studied using Hitachi-S4700 SEM. The test samples were prepared by ultrasonic wave mixing of CBs and CBs-N_2 with Di-water and allowed to set for 24 h (see Fig. 3.14). The results are shown in Fig. 3.15. The SEM images showed that the particle size of CBs became much smaller after N_2-plasma treatment than without any treatment. The SEM results proved that the plasma treatment method was an effective way to improve the dispersion of nanoscale powders.

3.3.2 *Particle-Size Analysis (PSA) of Nanoscale Powders*

Particle-size analysis (PSA), particle size measurement or simply particle sizing is the collective name of the technical procedures

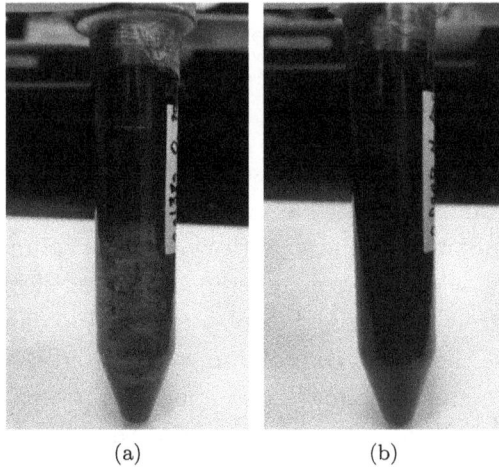

(a)   (b)

Fig. 3.14 Sample pictures of (a) CBs without plasma treatment and (b) CBs with N_2-plasma treatment.

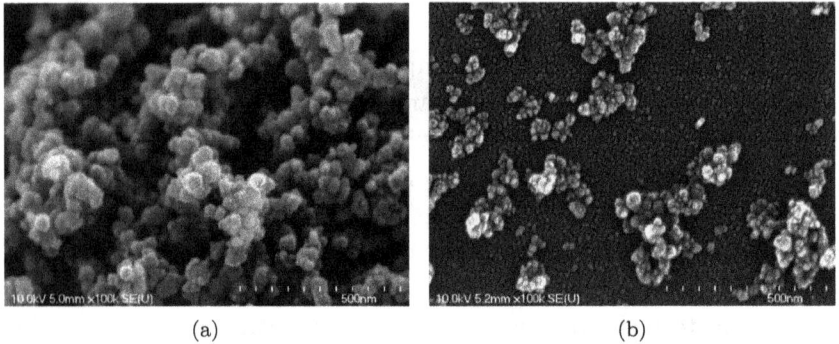

(a)   (b)

Fig. 3.15 SEM images of nanoscale powders with/without plasma treatment (a) CBs without plasma treatment and (b) CBs with N_2-plasma treatment.

or laboratory techniques. As the name suggests, it is a measuring method of the size distribution of individual particles in soil or liquid samples. It is a kind of effective means determining the size range and/or the average or mean size of the particles in a powder or liquid sample. The test data can be presented or used in some ways. The most common being a cumulative particle-size distribution curve, and we can affirm that PSA plays a key role in determining the bulk properties of the powder.

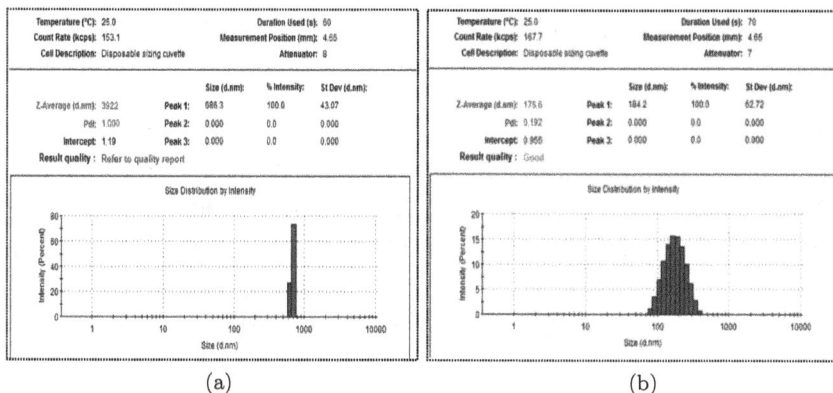

Temperature (°C): 25.0			Duration Used (s): 60
Count Rate (kcps): 153.1			Measurement Position (mm): 4.65
Cell Description: Disposable sizing cuvette			Attenuator: 8

	Size (d.nm):	% Intensity:	St Dev (d.nm):
Z-Average (d.nm): 3922 Peak 1:	688.3	100.0	43.07
Pdl: 1.000 Peak 2:	0.000	0.0	0.000
Intercept: 1.19 Peak 3:	0.000	0.0	0.000
Result quality : Refer to quality report			

Temperature (°C): 25.0			Duration Used (s): 70
Count Rate (kcps): 167.7			Measurement Position (mm): 4.65
Cell Description: Disposable sizing cuvette			Attenuator: 7

	Size (d.nm):	% Intensity:	St Dev (d.nm):
Z-Average (d.nm): 175.6 Peak 1:	184.2	100.0	62.72
Pdl: 0.192 Peak 2:	0.000	0.0	0.000
Intercept: 0.955 Peak 3:	0.000	0.0	0.000
Result quality : Good			

(a)   (b)

Fig. 3.16 PSA of nanoscale powders with/without plasma treatment (a) CBs without plasma treatment and (b) CBs with N_2-plasma treatment.

To further confirm the influence of plasma treatment processing on nanoscale powders, the PSA was introduced to study the size distribution of individual particles in CBs. All the test samples were diluted to the same concentration. The PSA data of samples A and B (similar to Fig. 3.14) were just as shown in Fig. 3.16(a)–(b). Figure 3.16(a) and (b) show the size distribution of individual particles of CBs without and with N_2-plasma treatment. By comparing the two sets of data, it could be found that the average particle size of CBs-N_2 became much smaller from 3922 nm to 175.6 nm than untreated CBs.

The PSA results proved that plasma treatment successfully reduced the particle size of powders. In other words, the plasma treatment method enhanced the dispersion of nanoscale powders. This conclusion is consistent with the previous SEM test conclusion.

3.3.3 *Atomic Force Microscope (AFM) of Nanoscale Powders*

Atomic force microscope (AFM) can be regarded as a kind of scanning probe microscopy (SPM), with demonstrated resolution on the order of fractions of a nanometer, more than 1,000 times better than the optical diffraction limit. The messages are gathered by "feeling" and "touching" the surface with a precision mechanical probe.

During the working process of AFM, the piezoelectric elements that facilitate tiny but accurate and precise movements on (electronic) command enable precise scanning. AFM has several advantages over the SEM. Unlike the electron microscope, which provides a two-dimensional projection or a two-dimensional image of a sample, the AFM provides a three-dimensional (3D) surface profile. Beyond that, samples viewed by AFM do not require any special treatments (such as metal/carbon coatings) that would irreversibly change or damage the sample and do not typically suffer from charging artifacts in the final image.

In this study, AFM was used to observe the stereoscopic structure of nanoscale particles, and then we could analyze the change of particle size. Finally, we could infer the change of dispersibility. Figures 3.17(a)–(b) show the 3D images of untreated CBs and CBs-N_2 particles. They compare the AFM images of the untreated particle (like Fig. 3.17(a)) with N_2 plasma-treated particles (like Fig. 3.17(b)). Through the 3D images and particle size information in the vertical dimension, it could be known whether the particle size of the particle with N_2 plasma treatment was much smaller than a

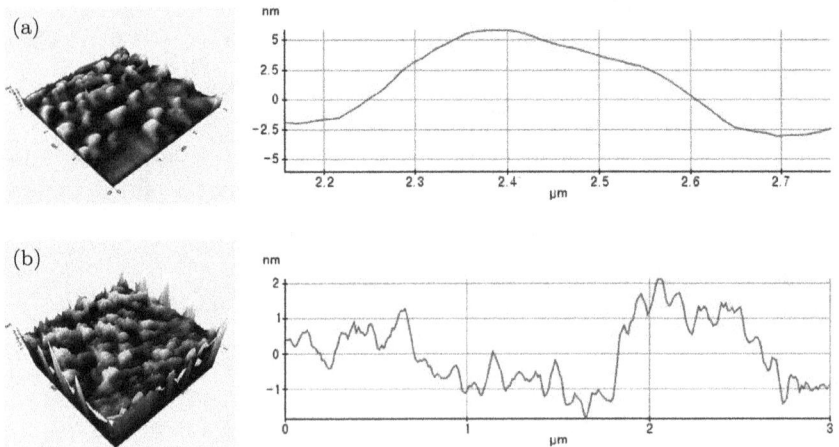

Fig. 3.17 AFM images of nanoscale powders with/without plasma treatment (a) CBs without plasma treatment (b) CBs with N_2 plasma treatment.

particle with untreated CBs. This point could reinforce the SEM and PSA conclusions.

3.3.4 *FTIR Spectrum of Nano-Scale Powders*

Fourier-transform infrared spectroscopy (FTIR) is a technique used to obtain an infrared spectrum of absorption or emission of a solid, liquid or gas. The FTIR spectrometer consists of a light source (silicon carbon rod, high-pressure mercury lamp), Michelson interferometer, detector and interferometer. FTIR spectrometer is a core part of the Michelson interferometer; place the samples in front of the detector, and the samples will absorb infrared light at certain frequencies. The interference light intensity received by the detector will change, and then interferogram of various samples will be obtained. FTIR identifies chemical bonds in a molecule by producing an infrared absorption spectrum. The spectra produce a profile of the sample, a distinctive molecular fingerprint that can be used to screen and scan samples for many different components. FTIR is an effective analytical instrument for detecting functional groups and characterizing covalent bonding information. By analyzing the changing of the FTIR absorbance peak, the changes of surface functional groups were analyzed. The FTIR analysis helps clients understand materials and products. Analytical testing sample screens, profiles and data interpretation are available globally from our experts who deploy FTIR to identify chemical compounds in consumer products, paints, polymers, coatings, pharmaceuticals, foods and other products. The FTIR spectra of nanoscale powders were measured with an FTIR-6300 type. The Origin 8.0 software was used to process FTIR data.

The FTIR curves are shown in Fig. 3.18. Figure 3.18 shows the FTIR spectra of CBs and CBs-N_2. It was obvious that the peak appeared near at $652\,cm^{-1}$, $880\,cm^{-1}$, $1049\,cm^{-1}$, $1541\,cm^{-1}$, $1700\,cm^{-1}$, 3393–$3926\,cm^{-1}$ in CBs without any treatment. They refer to $-C=CH_2$, $-CHO$, $C-O-C$, $RCOO-$, $RCOOH$, and $-OH$, respectively. This result was consistent with the CB's structural formula (see Fig. 3.8). Although most of the peaks that appeared

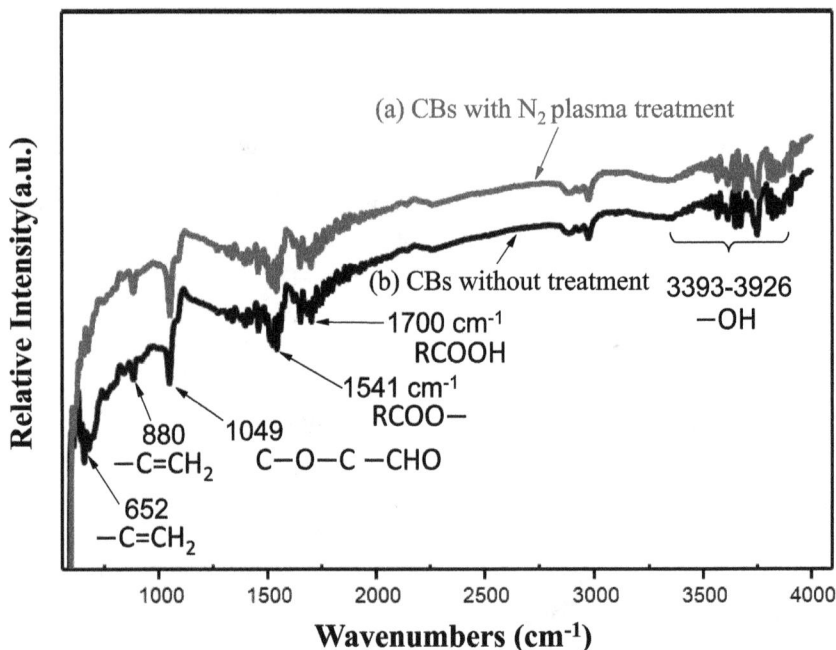

Fig. 3.18 FTIR of nanoscale powders (a) CBs without plasma treatment (b) CBs with N_2-plasma treatment.

in CBs without any treatment also appeared in CBs-N_2 simultaneously, minor differences still emerged. For instance, the peak at $652\,\mathrm{cm}^{-1}$, which refers to $-C=CH_2$, disappeared in CBs-N_2. However, the peak value at both $880\,\mathrm{cm}^{-1}$ and $1049\,\mathrm{cm}^{-1}$ had increased. This indicated that after plasma treatment, the carbon–carbon double bond ($-C=C-$) was cleaved and replaced by carbon–oxygen bonds ($-C-O-C-$, $-COO-$, $-C=O$). The oxygen is derived from Di-water or air. This explains why the dispersibility of CBs with plasma treatment in Di-water was enhanced, which is that the polar oxygen-containing groups were incorporated into the surface of the treated CB, which decreased the aggregation of CB particles and caused them to be more evenly disperse in the paste. Moreover, the tendency of aggregation of the treated CB particles was decreased, and the size of the particles became small.

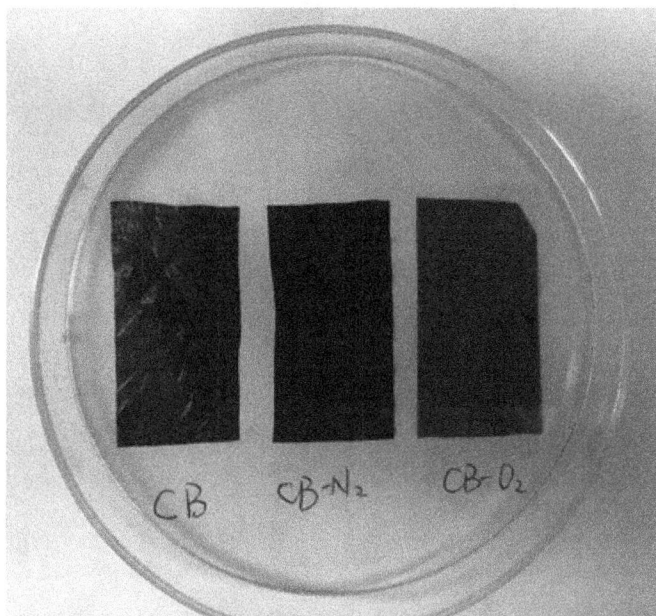

Fig. 3.19   Digital camera images of CBs/PVDF-HFP, CBs-N_2/PVDF-HFP and CBs-O_2/PVDF-HFP electrospun nanofibers.

These results further explained the SEM, PSA and AFM results in theory.

3.3.5   Morphologies of Complex Nanofibers

The surface morphologies of complex nanofibers without plasma treatment and with plasma treatment CBs were observed by using Hitachi S-4800 SEM at an accelerating voltage of 15 kV. Figure 3.19 presents the digital camera images of respective electrospun nanofibers. Their corresponding SEM images are shown in Fig. 3.20(a)–(f).

The surface morphology images of untreated-CBs/PVDF-HFP composite nanofibers were demonstrated in Fig. 3.20(a)–(b). It was observed that fibers were not smooth but somewhat rougher because of the existence of a small amount of CBs. However, a large number of CBs existed around the nanofibers in the form of particle

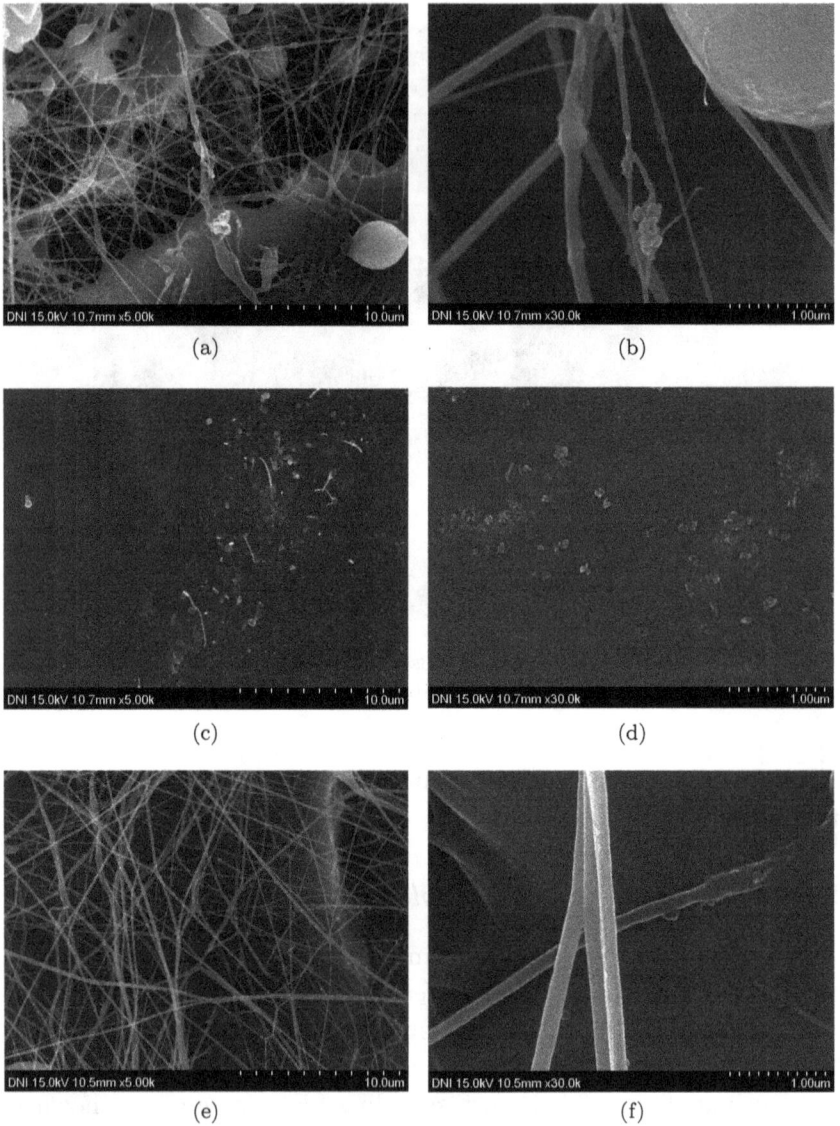

Fig. 3.20 SEM images of (a–b) CBs/PVDF-HFP, (c–d) CBs-N$_2$/PVDF-HFP and (e–f) CBs-O$_2$/PVDF-HFP electrospun nanofibers.

agglomeration. This indicates that untreated CBs cannot be well-dispersed in electrospun nanofibers. Figures 3.20(C)–(D) show the surface morphology of CBs-N$_2$/PVDF-HFP, and the images were much different from the images of CBs/PVDF-HFP.

The fibrous structure cannot be observed in the scanning electron micrographs of CBs-N$_2$/PVDF-HFP. Instead, a sheet of CBs was present in the SEM images. This might manifest the good interconnection between CB particles and PVDF-HFP fibers' surface. Figures 3.20(E)–(F) illustrated the configuration of the CBs-O$_2$/PVDF-HFP complex nanofibers surface. There was no obvious CB aggregates structure such as seen in Fig. 3.20(A) that stayed around the PVDF-HFP nanofibers, and regular fiber structures that could be discovered in Fig. 3.20(E). Figure 3.20(F) could assist to further observe the regular fiber structures. The images showed that the composite nanofibers were not perfectly smooth, and some non-uniform diameters of composite nanofibers were ascertained. It was mostly because of the protruded CB segments which were shrouded on the surface or within the inside of the composite nanofibers. Besides, it could be seen that the diameters of CBs-O$_2$/PVDF-HFP nanofibers were more uniform and larger than CBs/PVDF-HFP nanofibers. This might be because the dispersion of CBs-O$_2$ particles was better than CB particles.

In combination with all SEM images of complex nanofibers and referring to the Chapter 3 about the plasma principle, we can hypothesize about plasma-treated CB dispersion process, i.e., after high-energy plasma treatment, the surface characteristics of the CBs can be changed by non-polar and polar components, resulting in improvement in the surface polar functional groups and an increase in acidic and basic surface functional groups and ion-exchange properties.[18, 19] After N$_2$ plasma treatment, the electron-acceptor and electron-donor were formed on the surface of the CB particles, which further formed the electrical double-layer structure that played an important role in the interaction between the CBs and the matrix. However, for O$_2$ plasma-treated CBs, the active oxygen gas would react with most of the acidic and basic surface functional groups to form polar oxygen-containing groups, increasing the specific polar

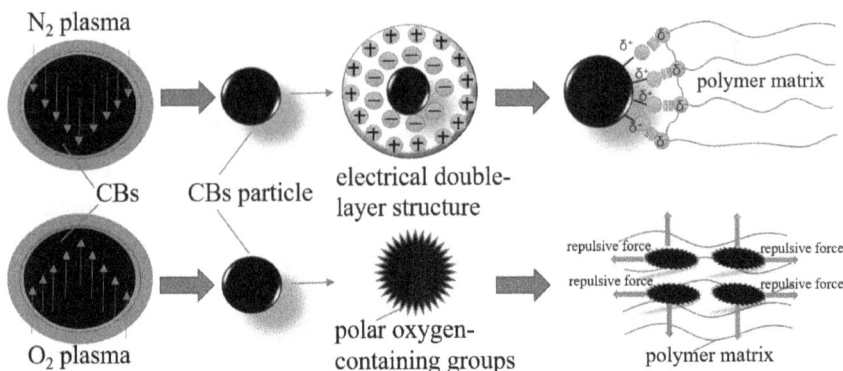

Fig. 3.21 The hypotheses of N_2 and O_2 plasma treating and dispersion processes.

component of the surface free energy of CB particles. This helped to increase the dispersibility of the CBs in the electrospun solution. Figure 3.21 could help to understand the hypotheses of N_2 and O_2 plasma treating and dispersion processes.

3.3.6 *Elemental Composition and Distribution Analysis of Complex Nanofibers*

Energy Dispersive Spectroscopy (EDX or EDS) analysis provides elemental and chemical analysis of a sample inside the SEM, TEM or FIB. It can be used to find the chemical composition of materials down to a spot size of a few microns and to create element composition maps over a much broader raster area.

The EDX mapping and spectra of tissue sections were performed on the same SEM/EDX system described earlier. The results were detected by Horiba EMAX silicon drift X-ray detector. By identifying the EDX imaging, we attempted to correlate the elemental composition measured across nanofibers to representatives of different dispersion degrees of untreated-CBs and plasma-treated CBs. All the EDX analyses were carried out in the same conditions where operating voltage was 15 kV, WD was 12 mm, scan size was 1024 px and ICR was 9.9 kcps. Figures 3.22(a–c) show images of EDX analysis. The carbon and fluorine elementals were mainly measured

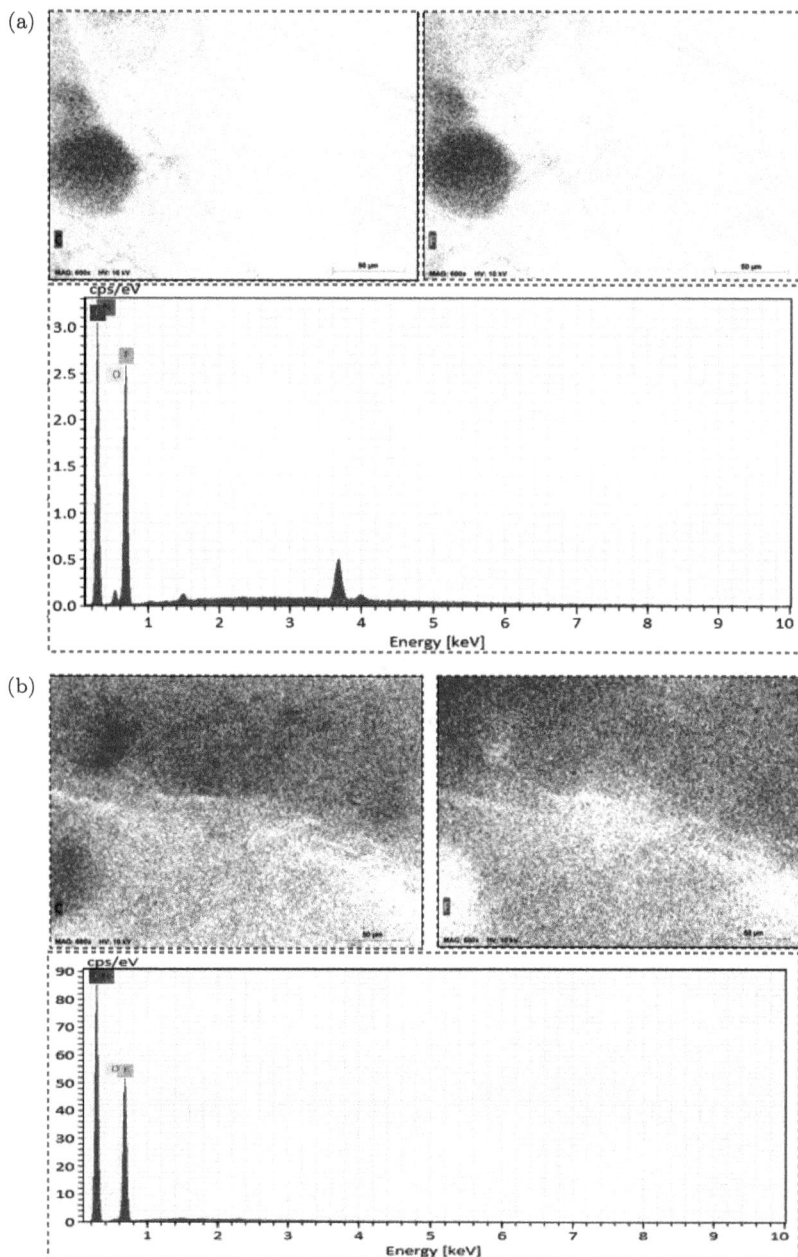

Fig. 3.22 EDX spectra and mapping of (a) CBs/PVDF-HFP, (b) CBs-N$_2$/PVDF-HFP, (c) CBs-O$_2$/PVDF-HFP electrospun nanofibers.

(c)

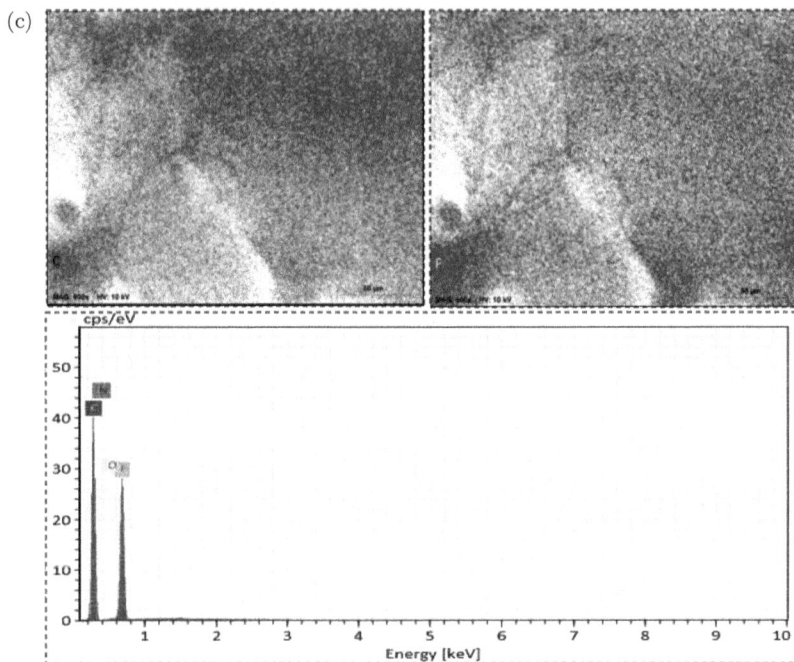

Fig. 3.22 (*Continued*)

Table 3.2 The element mass ratios on certain area surfaces of CBs/PVDF-HFP nanofibers

Element	At No.	Mass (%)	Mass Norm. (%)	Atom
Carbon	6	51.30	51.30	61.97
Nitrogen	7	1.02	1.02	1.06
Oxygen	8	3.85	3.85	3.5
Fluorine	9	43.83	43.83	33.47

in this test. EDX spectra results indicated the equal size layers of all three kinds of nanofibers, containing carbon (C) and fluorine (F) elementals. Tables 3.2–3.4 show detailed information about the content ratio of carbon and fluorine of the three kinds of nanofibers. The content ratio of the two kinds of elementals was a little different. In CBs-N_2/PVDF-HFP nanofibers, carbon content was maximum. It suggests that CBs-N_2 was dispersed very well on PVDF-HFP

Table 3.3 The element mass ratios on certain area surfaces of CBs-N_2/PVDF-HFP nanofibers

Element	At No.	Mass (%)	Mass Norm. (%)	Atom
Carbon	6	60.09	60.09	70.27
Nitrogen	7	0.46	0.46	0.46
Oxygen	8	0.76	0.76	0.66
Fluorine	9	38.69	38.69	28.60

Table 3.4 The element mass ratios on certain area surfaces of CBs-O_2/PVDF-HFP nanofibers

Element	At No.	Mass (%)	Mass Norm. (%)	Atom
Carbon	6	57.57	57.57	68.07
Nitrogen	7	0.58	0.58	0.59
Oxygen	8	0.69	0.69	0.61
Fluorine	9	41.15	41.15	30.75

nanofibers or have good interfacial interaction between the CBs and polymer matrix. However, the content ratio of carbon and fluorine was almost the same on the equal size layers of both untreated-CBs/PVDF-HFP and CBs-O_2/PVDF-HFP nanofibers.

Nonetheless, judged from the EDX mapping (distribution of element concentration values), it was observed that the dispersion degree of CBs-O_2 on PVDF-HFP polymer was better than pure CBs. The reason why the carbon content in CBs-O_2/PVDF-HFP nanofibers saw no significant increase as compared with CBs/PVDF-HFP is maybe because most of the CBs-O_2 are dispersed inside the fiber and are hard to detect. The conclusions which were obtained by the EDX analysis could help further prove the hypotheses proposed before.

3.3.7 *Surface Chemical Bonding States of Complex Nanofibers*

FTIR is a technique used to obtain an infrared spectrum of absorption or emission of a solid, liquid or gas. The FTIR spectra of

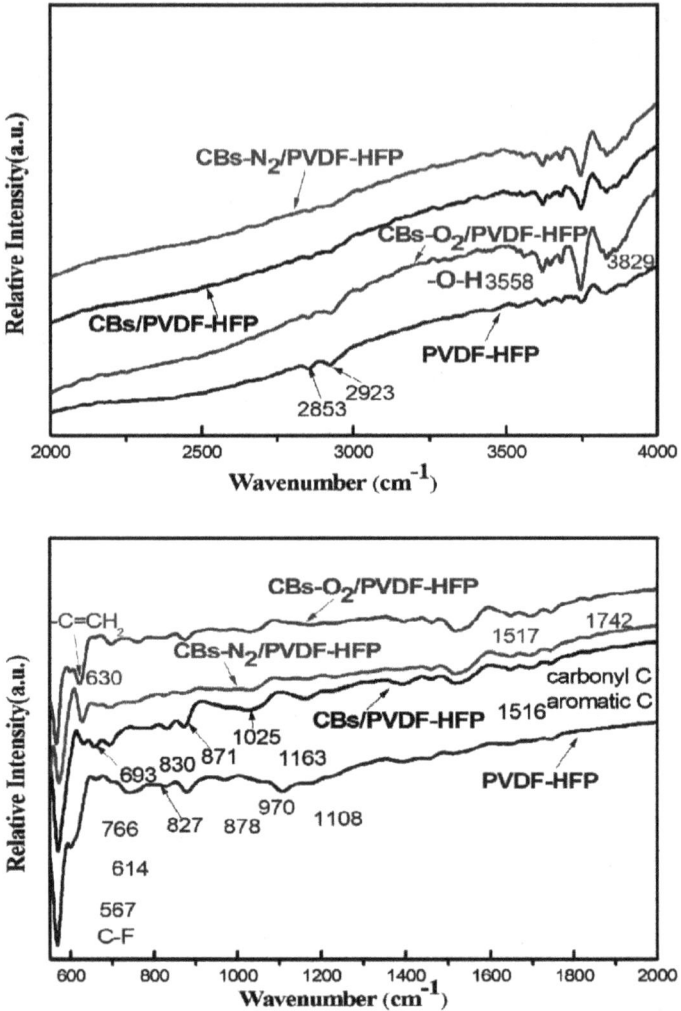

Fig. 3.23 IR spectra of PVDF-HFP, CBs/PVDF-HFP, CBs-N₂/PVDF-HFP, CBs-O₂/PVDF-HFP electrospun nanofibers.

electrospun composite nanofibers were measured with an FTIR-6300 type A instrument in the transmission mode at room temperature in the range of $550\,\mathrm{cm}^{-1}$–$4000\,\mathrm{cm}^{-1}$. The results are shown in Fig. 3.23. In the FTIR spectrum of pure PVDF-HFP, the corresponding crystalline peak was observed at $567\,\mathrm{cm}^{-1}$, representing the bending vibrations of the C–F group. The peak at $878\,cm^{-1}$ was attributed to the amorphous phase of PVDF-HFP and could not be used to identify any of the crystalline phases.[20] The bands at $614\,cm^{-1}$ (CF$_2$ stretching), $766\,cm^{-1}$ (CF$_2$ stretching) and $970\,cm^{-1}$ (CH stretching) were associated with α-phase characteristic peaks of PVDF-HFP.[21,22] The band at $614\,cm^{-1}$ showed a mixed-mode of C–C–C skeletal vibration and CF$_2$ vibration.[21,23] The peak at $1108\,cm^{-1}$ suggested the formation of β-phase (CF$_2$ stretching).[24] The band that appeared at $1108\,cm^{-1}$ was mainly formed by CF$_2$ symmetric stretching mode.[25] The CH$_2$ group was described in two frequencies between $2800\,cm^{-1}$ and $3000\,cm^{-1}$.[21] The FTIR spectra of complex nanofibers were much different from pure PVDF-HFP. In the FTIR spectra of CBs/PVDF-HFP, such kind of α-phase peaks, which appeared in pure PVDF-HFP, decreased dramatically. It might indicate that the α-phase was transformed to β-phase (at $871\,cm^{-1}$), and γ-phase (frequently appearing as a shoulder, could be detected at $830\,cm^{-1}$ nearby $831\,cm^{-1}$ of CBs/PVDF-HFP).[20,26] The absorption peaks at $1516\,cm^{-1}$–$1742\,cm^{-1}$ (carbonyl and aromatic C) and $3558\,cm^{-1}$–$3829\,cm^{-1}$ (–OH stretching) in CBs/PVDF-HFP, CBs-N$_2$/PVDF-HFP, CBs-O$_2$/PVDF-HFP were the characteristic peaks of CBs complex nanofibers. At $878\,cm^{-1}$, the bands that referred to the amorphous phase of pure PVDF-HFP decreased in CBs-N$_2$/PVDF-HFP and CBs-O$_2$/PVDF-HFP. It can even hardly be found in CBs-N$_2$/PVDF-HFP. This means the plasma-treated CBs, especially N$_2$ plasma-treated CBs, can react with PVDF-HFP and help increase the crystallinity of PVDF-HFP. Also, the peak at $630\,cm^{-1}$ in CBs-N$_2$/PVDF-HFP and CBs-O$_2$/PVDF-HFP, was assigned to C=CH$_2$ stretching vibrations of CBs-N$_2$/PVDF-HFP and CBs-O$_2$/PVDF-HFP complex nanofibers. It indicated that the interconnection was formed between plasma-treated CBs and PVDF-HFP. Beyond this, the α-phase and β-phase were hard to find in FTIR of plasma-treated CBs and PVDF-HFP. The reason was that the generated –C=CH$_2$ banding could impact on the vibration of CF$_2$ dipole. Finally, after comparing CBs-N$_2$/PVDF-HFP with CBs-O$_2$/PVDF-HFP FTIR results, it could be seen that the peak value of hydroxyl groups (at $3558\,cm^{-1}$ to $3829\,cm^{-1}$) in CBs-O$_2$/PVDF-HFP was maximum. The reason was that during the plasma treatment process, O_2 plasma was used to add oxygen-containing polar functional groups. However, N_2 plasma was expected to improve the electron acceptor and an electron donor.[12]

3.3.8 *The Mechanical Behavior of Electrospun Complex Nanofibers*

The traditional stress-strain methods suitable for testing a series of electrospun nanofibers can be used to evaluate the mechanical properties of the electrospun nanofiber sheet of CBs/PVDF-HFP composite. Each sample used in the test was prepared in $4 \times 2\,cm^2$ size (Fig. 3.24). The thickness of composite nanofibers was measured before each test with a micrometer thickness gauge. The test was carried out in the device, as shown in Fig. 3.25. The stress-strain curves of nanofibers were drawn based on the average of the five sets of test data. As it can be observed in Fig. 3.26, even though Young's modulus (the slope of the curves) of CBs-N_2/PVDF-HFP was almost the same as Young's modulus of CBs/PVDF-HFP, both tensile strength and strain at break were at maximum value in these three kinds of electrospun nanofibers. The tensile strength of

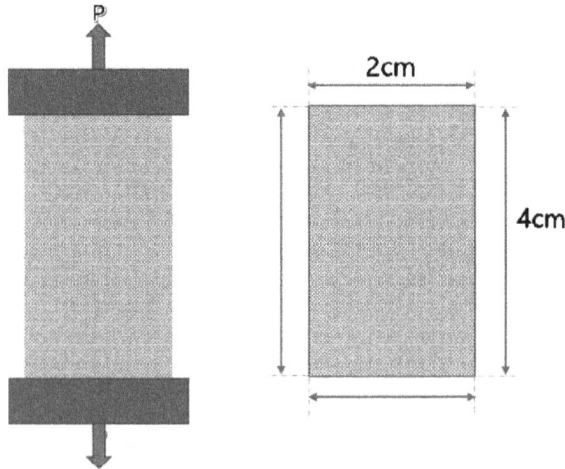

Fig. 3.24 Schematic diagram of stretch tension sample.

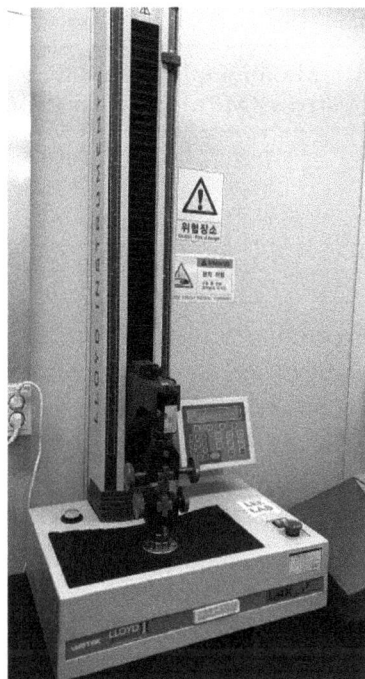

Fig. 3.25   Tension testing machine.

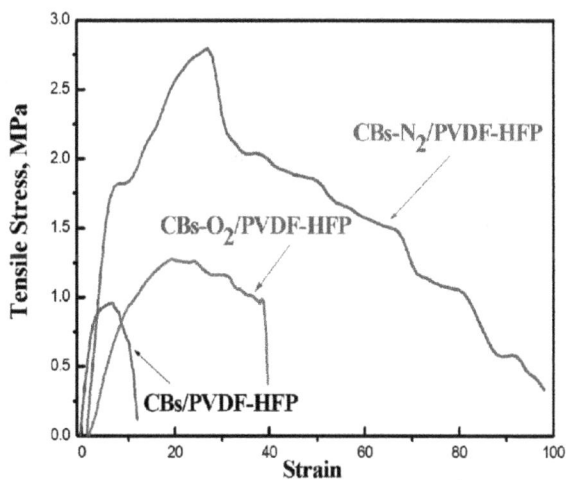

Fig. 3.26   Stress-strain curves of CBs/PVDF-HFP, CBs-N_2/PVDF-HFP, CBs-O_2/PVDF-HFP composite nanofibers.

Table 3.5 The maximum of stress-strain curves

Material	Maximum stress (MPa)	Maximum strain (%)	Modulus (MPa)
CB-N$_2$/PVDF-HFP	2.7986	27.0000	0.1037
CB-O$_2$/PVDF-HFP	1.2748	19.2846	0.0661
CB-PVDF-HFP	0.9599	6.5027	0.1476

CBs-N$_2$/PVDF-HFP complex nanofibers was 2.7986 MPa, which was nearly three times higher than the 0.9599 MPa of CBs/PVDF-HFP complex nanofibers. The strain value at break decreased from 27% for CBs-N$_2$/PVDF-HFP to 6.5% for CBs/PVDF-HFP. The detailed data were listed in Table 3.5. Therefore, the CBs-N$_2$/PVDF-HFP electrospun complex nanofibers had a higher stiffness, strength and ductility than the other two nanofibers. The enhancement in the mechanical properties might originate from the good dispersion of CBs-N$_2$ and the strong interaction between CBs and the PVDF-HFP polymer matrix. Since N$_2$ plasma treatment led to an increase in a specific component, it increased acid–base values and ion exchange of the carbon surfaces.[19] However, the tensile strength and strain value at the break of CBs-O$_2$/PVDF-HFP were 1.2748 MPa and 19.2846%, respectively. Although these two values were much bigger than the two CBs/PVDF-HFP, Young's modulus decreased. The reason might be that O$_2$ plasma treatment was used to add the polar functional groups, which dramatically increased the dispersibility through increasing specific polar component of the surface-free energy of CBs.[19] This results in good dispersibility but poor interaction between CBs and polymer matrix. The findings were consistent with SEM and FTIR analyses.

3.4 Conclusions

In this research, first, the plasma was used to treat the nanoscale particles, and the dispersion of nanoscale particles was studied. The results showed that plasma treatment was an effective way to decrease the aggregation of nanoscale particles and improve the dispersibility of nanoscale particles in Di-water. The principle of increasing dispersion by plasma was that the polar oxygen-containing groups were incorporated into the surface of treated nanoscale particles. Then, the pure-CBs/PVDF-HFP, plasma-treated CBs/PVDF-HFP complex nanofibers were fabricated by the electrospinning process. N_2 and O_2 were used as plasma treatment sources. The mechanical behavior of pure-CBs and plasma-treated CBs electrospun complex nanofibers were studied. According to the SEM and EDX results, it can be preliminaryily judged that the dispersibility of plasma-treated CBs in electrospun complex nanofibers was better than of pure-CBs. The FTIR spectra were used to study the crystal structure of complex nanofibers. The results showed two possibilities. The first point was that the interconnection ($-C=CH_2$ banding), which would impact the vibration of CF_2 dipole, was formed between plasma-treated CBs, especially N_2 plasma-treated CBs and PVDF-HFP. This help increased the crystallinity of PVDF-HFP. The second point was that O_2 plasma treatment was used to add oxygen-containing polar functional groups to improve CBs' dispersibility. However, N_2 plasma treatment was expected to improve the electron acceptor and electron donor of CBs to achieve a good connection with the polymer matrix. These two points further declare that plasma treatment is an effective method to improve CB dispersibility in electrospun nanofibers and enhance interfacial interaction between the CBs and polymer matrix by increasing specific polar component or including an electron acceptor and an electron donor. Finally, the mechanical behavior test results showed that both CBs-N_2/PVDF-HFP and CBs-O_2/PVDF-HFP have a higher strength and ductility than pure-CBs/PVDF-HFP electrospun complex nanofibers. It could result in good dispersibility and good interfacial interaction. Besides, on comparing the stress-strain curves of CBs-N_2/PVDF-HFP and CBs-O_2/PVDF-HFP, it can be seen that strong interfacial interaction was more effective than good dispersibility for improving the mechanical behavior of complex materials. Based on this research, the plasma process can be further exploited in treating carbon nanoparticles for improving the performance of electrospun complex fibers.

References

[1] F. El-Tantawy, K. Kamada and H. Ohnabe, *Mater. Lett.*, 56(1–2), 2002, 112–126.

[2] A. I. Medalia, *Rubber Chem. Technol.*, 59(3), 1986, 432–454.

[3] S. G. Chen, J. W., Hu, M. Q. Zhang, M. Z. Rong and Q. Zheng, *Sensor. Actuat. B: Chem.*, 113(1), 2006, 361–369.

[4] R. Richner, S. Müller and A. Wokaun, *Carbon*, 40(3), 2002, 307–314.

[5] D. M. Bigg and D. E. Stutz, *Polym. Compos.*, 4(1), 1983, 40–46.

[6] J. Hwang, J. Muth and T. Ghosh, *J. Appl. Polym. Sci.*, 104(4), 2007, 2410–2417.

[7] Y. Lin, T. W. Smith and P. Alexandridis, *Langmuir*, 18(16), 2002, 6147–6158.

[8] F. Tiarks, K. Landfester and M. Antonietti, *Macromol. Chem. Phys.*, 202(1), 2001, 51–60.

[9] N. Tsubokawa, *Progr. Polym. Sci.*, 17(3), 1992, 417–470.

[10] P. E. Fanning and M. A. Vannice, *Carbon*, 31(5), 1993, 721–730.

[11] C. Chen, A. Ogino, X. Wang and M. Nagatsu, *Appl. Phys. Lett.*, 96(13), 2010, 131504.

[12] R. J. Goldston and P. H. Rutherford, *Introduction to Plasma Physics*, CRC Press, 1995.

[13] A. I. Morozov, *Introduction to Plasma Dynamics*, CRC Press, 2013.

[14] A. Ziabicki, *Fundamentals of Fibre Formation: The Science of Fibre Spinning and Drawing*, Wiley, 1976.

[15] D. Li and Y. Xia, *Adv. Mater.*, 16(14), 2004, 1151–1170.

[16] J.-B. Donnet, R. C. Bansal and M.-J. Wang, *Carbon Black: Science and Technology* (2nd ed.), Routledge, 2018.

[17] H. Bipp and H. Kieczka, Formamides. In *Ullmann's Encyclopedia of Industrial Chemistry*. Wiley, 2011.

[18] G. Akovali and I. Ulkem, *Polymer*, 40(26), 1999, 7417–7422.

[19] S.-J. Park and J.-S. Kim, *J. Colloid Interface Sci.*, 244(2), 2001, 336–341.

[20] Q. Meng, W. Li, Y. Zheng and Z. Zhang, *J. Appl. Polym. Sci.*, 116(5), 2010, 2674–2684.

[21] N. Betz, A. Le Moël, E. Balanzat, J. M. Ramillon, J. Lamotte, J. P. Gallas and G. Jaskierowicz, *J. Polym. Sci. Part B: Polym. Phys.*, 32(8), 1994, 1493–1502.

[22] P. Martins, A. C. Lopes and S. Lanceros-Mendez, *Progr. Polym. Sci.*, 39(4), 2014, 683–706.

[23] B. Mattsson, H. Ericson, L. M. Torell and F. Sundholm, *J. Polym. Sci. Part A: Polym Chem.*, 37(16), 1999, 3317–3327.

[24] J. C. Li, C. L. Wang, W. L. Zhong, P. L. Zhang, Q. H. Wangand and J. F. Webb, *Appl. Phys. Lett.*, 81(12), 2002, 2223–2225.

[25] V. Bharti, T. Kaura and R. Nath, *IEEE Trans. Dielectr. Electr. Insul.*, 4(6), 1997, 738–741.

[26] Y. Bormashenko, R. Pogreb, O. Stanevsky and E. Bormashenko, *Polym. Test.*, 23(7), 2004, 791–796.

Part 3
Fracture Mechanics for Advanced Composites: Polymer Composites, Laminated Composites and Nanocomposites

Chapter 4

Characteristics of Up-Cycling Fibers Using Slag: Fiberization Process, Mechanical Properties

Chang-Wook Park[*,§], Se-Yoon Kim[†,¶] and Yun-Hae Kim[‡,‖]

*Ocean ICT & Advanced Materials
Technology Research Division,
Research Institute of Medium & Small Shipbuilding,
Busan, Republic of Korea
†Major of Materials Engineering,
Department of Marine Equipment Engineering,
Korea Maritime and Ocean University,
Busan, Republic of Korea
‡Department of Ocean Advanced Materials
Convergence Engineering,
Korea Maritime and Ocean University,
Busan, Republic of Korea
§pcw0591@naver.com
¶seyun8269@naver.com
‖yunheak@kmou.ac.kr

Slag can be divided into large blast furnace slag and steel slag, accounting for more than 470 million tons of emissions in Korea. Much of the blast furnace slag is a trend that recycles material for cement and is recognized as a new resource from waste. This material can replace the currently used industrial fibers (glass, basalt and carbon fibers). Slag fibers have economic and environmental benefits because of the waste recycling. Fibers produced by mixing various slags have a chemical composition similar to that of basalt fibers. It is anticipated that this can be treated as alternative fibers to basalt fibers and can help overcome price competitiveness while protecting natural resources.

4.1 Introduction

Domestic steel-making slag produced amounts to about 20 million tons or more annually. Generally, in the process of producing 1 ton of steel, 100–500 kg of steel-making slag is generated. For example, Fe–Ni slag generated per year amounts to more than 1 million tons and fly ash, which is an industrial by-product, amounts to about 1.3 million tons annually, based on one thermal power plant. The generated slag is dumped into the yard, and it cools and coagulates in the atmosphere with a large amount being waterproofed. The solidified clay of slag is recycled as a raw material for cement, but some are buried and processed due to market conditions. However, it is unconducive to create high value-added product or use as a raw material. Therefore, there is a need for a new method, different from the application to low value-added industries, and the recovery of metals or resources from existing single-composition waste resources. Thus, a new recycling method could be presented through the combination of various process by-products.

Representative industrial fibers such as carbon fiber, glass fiber and basalt fiber have been applied to various industrial fields. Carbon fibers have many advantages such as tensile strength, elasticity and corrosion resistance, but their disadvantage is that they are expensive. Glass fiber is applied to many industrial fields because of its cheap price, but it causes serious unhealthy work environment and health problems. So, much effort is being made to replace them. Basalt fibers have been developed for the replacement of glass fibers. Currently, much research is underway to replace glass fibers, but basalt fiber raw materials are undervalued compared to glass fibers in terms of their originality as natural resources and low competitiveness in price. Fibers produced by mixing various slags have a chemical composition similar to that of basalt fibers. It is anticipated that these can be used as alternative fibers for basalt fibers as they can overcome price competitiveness while protecting natural resources.

In the following sections, we will study the applicability of slag fiber in industries through the analysis of their characteristics, spinning processes and fiber properties.

4.2 Raw Materials

The composition ratio and characteristics depend on the type of slag. In order to make a long fiber by mixing various kinds of slags, proper composition ratio, proper mixing ratio and possibility of fiber spinning are required as seen through the analysis of raw materials. In general, not many types of rocks are suitable for high strength and high resistance to heat, acid alkali and moisture. The most representative rock is basalt, with SiO2 47–56%, Al_2O_3 14–19%, $FeO + Fe_2O_3$ 7–15%, P_2O_5 0.3–0.8%, Cr_2O_3 0.04%, $K_2O + Na_2O$ 2.5–6%, CaO 8–11% and MO 8–11%.

Although the composition ratio of various types of slags is not completely constant, the slag removal process produces a certain percentage of the slag. Fly ash consists of more than 80% of SiO_2 and Al_2O_3, Fe–Ni slag accounts for more than 80% of SiO_2 and MgO, electric furnace slag accounts for more than 50% of CaO and Fe_2O_3, and Tallin slag is a slag made up mainly of more than 50% of CaO and Fe_2O_3, with the possibility of combining each slag at an appropriate ratio. In addition, the analysis of raw materials shows the possibility and conditions of fiberization.

In this section, the analysis of raw materials was conducted to study whether or not the fibers were spun using slag. Raw material analysis was undertaken for the continuous fiber by mixing electric furnace slag, fly ash and Fe–Ni slag. In addition, the spinning of slag fibers was recognized through the analysis of raw materials of basalt fibers which are most similar to slag fibers. The mix of slag is represented in Table 4.1 by a three-way system.

Table 4.1 Mixing ratio of basalt and slag

Mixing ratio	Fly ash (%)	Furnace slag (%)	Fe–Ni slag (%)	Basalt (%)
Case 1	0	0	0	100
Case 2	40	30	30	0
Case 3	50	25	25	0
Case 4	60	20	20	0

4.2.1   Result of XRF Analysis

The results of chemical composition analysis according to basalt, individual slag and mixed slag are presented in Table 4.2 with the five elements having a significant effect on spinning. Specimens were quantified to 0.5 g of dried raw material at a temperature of 100°C after washing.

The composition of the compounded slag closely resembles that of basalt. Therefore, the ideal composition of the raw material for spinning would require compositing the chemical compounds similarly to that of the basalt fiber. Table 4.3 shows the components that affect the fiberization of basalt compounds for spinning continuous fiber.

According to the analysis, fly ash showed that SiO_2 and Al_2O_3 accounted for nearly 83% of the total, the furnace slag was mainly composed of $FeO + Fe_2O_3$ and CaO, and Fe–Ni slag was mostly composed of SiO_2 and MgO. Compared to basalt, slag had a moderate composition. The slag, which was mixed in three cases, also had a chemical composition ratio that was expected to enable continuous fiber spinning. The higher the content of fly ash, the greater the content of SiO_2, while the lesser the MgO.

Table 4.2   The slag compounds using XRF analysis

	SiO_2 (%)	Al_2O_3 (%)	$FeO + Fe_2O$ (%)	CaO (%)	MgO (%)
Basalt	50.43	15.57	12.61	8.78	8.79
Furnace slag	15.41	3.02	23.34	36.45	5.03
Fe–Ni slag	55.50	2.60	7.99	0.55	30.70
Fly ash	61.68	21.89	4.72	4.45	1.29
Slag (40:30:30)	45.70	12.81	15.21	9.52	13.90
Slag (50:25:25)	51.49	17.47	11.21	8.28	6.49
Slag (60:25:25)	54.68	17.56	9.48	8.15	3.48

Table 4.3   The fiberization of basalt compounds for spinning continuous fiber

SiO_2 (%)	Al_2O_3 (%)	$FeO + Fe_2O$ (%)	CaO (%)	MgO (%)
47–56	14–19	7–15	8–11	3.5–10

Acid coefficients are an important requirement in the analysis of raw materials for spinning. The higher the acid factor, the higher the viscosity of the melt and the lower the viscosity. The acid factor is calculated as $(SiO_2 + Al_2O_3)/(MgO + CaO)$. The acidity factor of basalt formation suitable for the manufacture of roving company will require an acid factor of 2.9–6.5. Slag containing 40% fly ash contains 2.5 and 50%, slag containing 4.67 and 60%, and the acid factor is 6.21; thus, the acidity factor is satisfied.

4.2.2   Result of XRD

The amorphous status of raw materials was verified through XRD analysis. Figure 4.1(a) is difficult to release because the temperature drops slowly over several hours at 1,450–1,500°C when the raw material is melted and cooled in the furnace. Cooling in the air in the RT condition after melting causes the temperature of the melt to drop quickly and cool rapidly, but there is a partial crystallization inside the melt. The larger the amount of raw materials, the more crystalline the parts, so continuous spinning is difficult. Continuous spinning in the non-deterministic part results in a short-circuit of the fiber in the crystalline part. Figure 4.1(b) can be seen visually when cooled in the air. Water after melting the raw materials will cool down quickly, so the temperature inside can drop rapidly to obtain a glass material suitable for the prevention of continuous fibers.[14]

(a)   (b)   (c)

Fig. 4.1   Classification of Slag depending on the different cooling condition in (a) furnace, (b) air and (c) water.

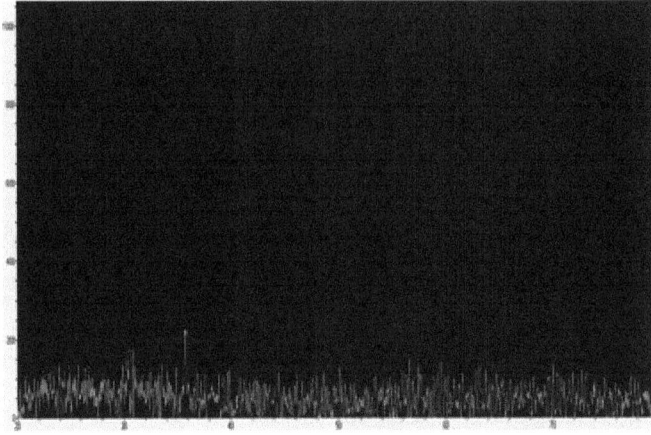

Fig. 4.2 XRD analysis on slag.

The state could be visually determined, and the presence or absence of static or abnormal quality is shown in Fig. 4.2 through XRD. In the case of water-cooled raw materials, all of them cannot reach the peak, such as in the XRD results. All raw materials, except water and air-cooled, failed in spinning because it is difficult to find spinning temperatures.

4.2.3 *Result of TG/DTA*

TG/DTA analysis was performed for the analysis of properties through the thermal decomposition behavior of slag and basalts of various types. It is possible to predict the design and spinning temperature of the fiber emitter by analyzing the high temperature characteristics of the raw material.[15] The experimental conditions were carried out in accordance with Table 4.4.

Figure 4.3 is an analysis of the pyrolysis behavior of the basalts. A steady decline in mass occurred without a clear inflection point. According to the DTA analysis, the graph adaptation is estimated to have a melting point based on the absorption peak at 1197.9°C, and the enthalpy is steadily increasing with only a change in the increase in the width of each peak. Glass transfer temperature indicates the temperature at the intersection of the tangent at the

Table 4.4 Condition of TG/DTA analysis

Conditions	
Atmosphere	Air
Specimen weight	0.5 g
Heating rate	10°C/min
Maximum temperature	1475°C

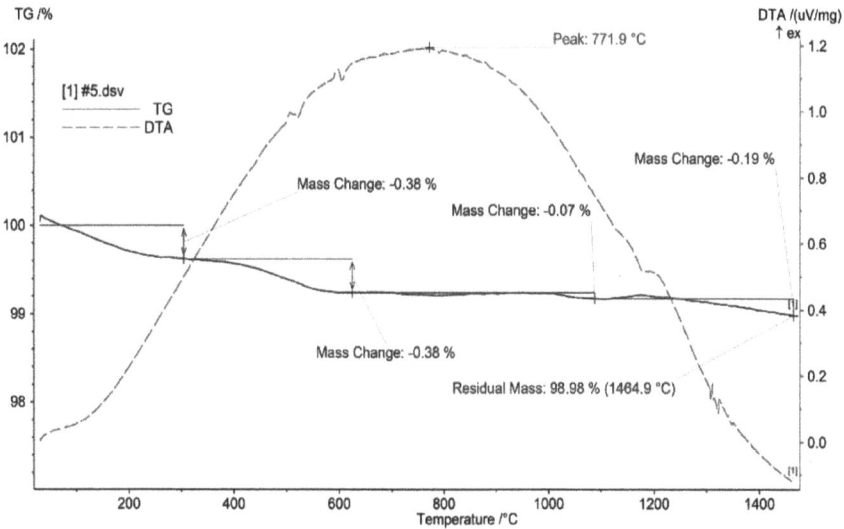

Fig. 4.3 TG/DTA data of basalt.

crystallization temperature of the DTA graph and the tangent at the fusion temperature. The crystallization temperature is about 468°C, the fusion point is 1197.9°C and the glass ion is about 877°C.

Figure 4.4 is a thermolysis behavior graph of a compound slag whose ratio of the fly ash is as follows: Furnace Slag:Fe–Ni Slag is 40:30. Heat peaks are formed at 783.4°C and mass-increasing adaptations caused by oxidation have resulted in a heating process throughout the entire process up to 1,160.2°C, with no crystallization or vitrification taking place. The absorption reaction at 1,323°C is

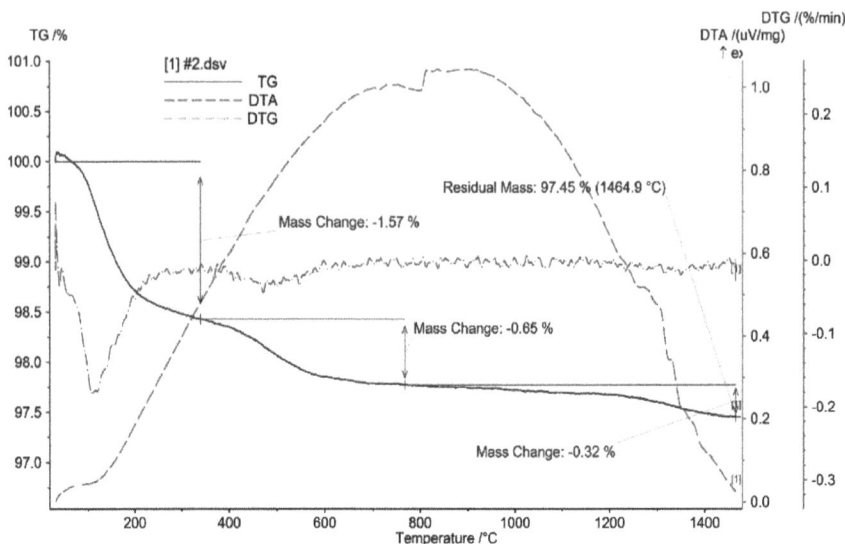

Fig. 4.4 TG/DTA data of 40:30:30.

estimated to be the melting point. The crystallization temperature is approximately 406°C, the melting point is 1323°C, and the glass ion is achieved approximately at 838°C.

Figure 4.5 is a thermolysis behavior graph of a compound slag whose ratio of fly ash is as follows: Furnace Slag:Fe–Ni Slag is 50:25:35. The mass reduction started at 30.44°C and stopped at 1,464°C. The crystallization of the specimen begins nearly at 310°C, the melting point of the specimen is around 1,085°C, the melting temperature is about 310°C, and the glass transfer temperature is 757°C.

Figure 4.6 is a thermolysis behavior graph of the mixed slag with a ratio of 60:20:20 of the fly ash:furnace slag:Fe–Ni slag. With the start of crystallization at the onset point of about 337°C, the dehydration reaction ended and a 1.57% mass reduction occurred. After that, the first gasoline process was carried out to the midpoint of about 734°C, and the gasoline process was completed and melted to the end set of 1,257°C. The crystallization temperature is about

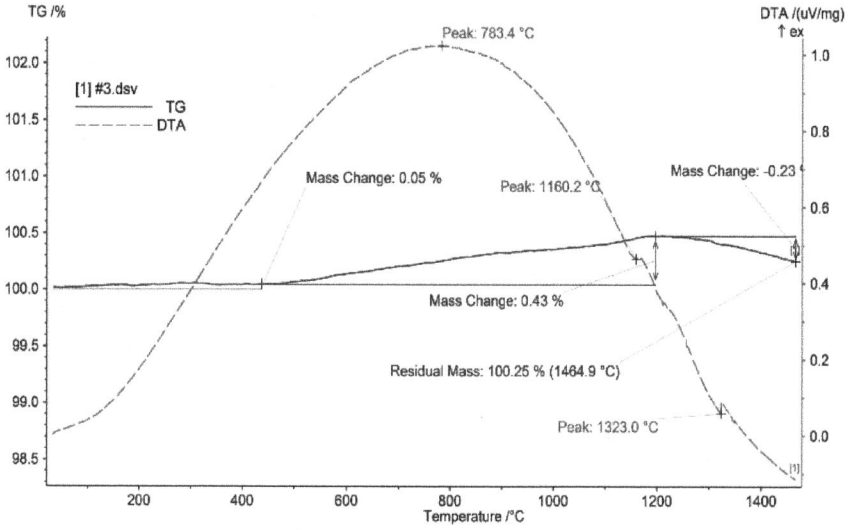

Fig. 4.5   TG/DTA data of 50:25:25.

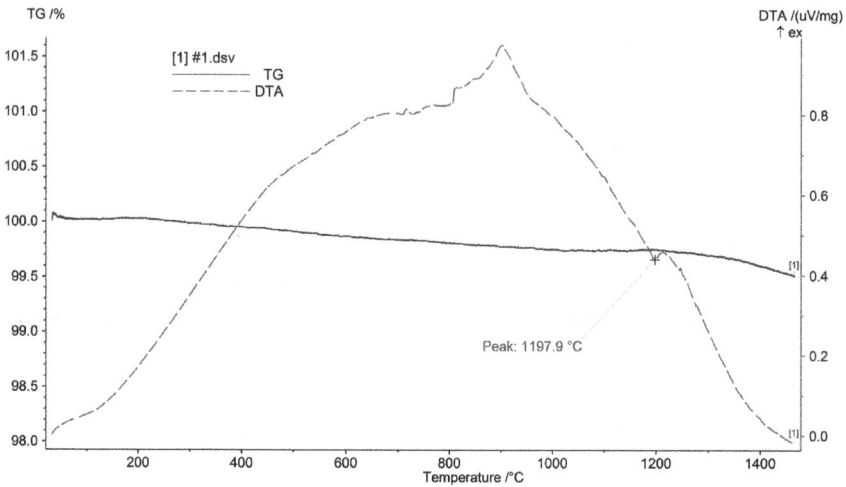

Fig. 4.6   TG/DTA data of 50:25:25.

Table 4.5   Main temperature of basalt and slag

	Crystallization temperature	Melting temperature	Glass transition temperature
Basalt	468.82	1,197.9	877.82
Slag (40:30:30)	406.67	1,323.0	838.67
Slag (50:25:25)	310.94	1,085.93	757.74
Slag (60:20:20)	337.44	1,257.94	812.94

337°C, the melting temperature is 1,197°C and the glass field temperature is 877°C.

Table 4.5 shows the temperature of crystallization, melting and glazing, respectively. The melting point and glass field showed no constant relationship with the mixing ratio of multi-species slag. The emitter was designed through the analysis of high-temperature thermal decomposition behavior of the slag. For samples with the highest melting point, the temperature of the furnace of the emitter should be at least 1,320°C.

4.2.4 Conclusion

In this section, raw material analysis was made to spin continuous fiber by mixing slag. Further, the analysis of basalt examined the possibility of the slag fiber being spun.

(1) Based on the chemical composition ratio of basalt fibers suitable for roving companies, we saw the possibility of spinning the combined slag. The higher the content of fly ash, the higher the content of SiO_2, while the content of MgO decreased, increasing the acid factor. However, all of the combined slag met all the criteria for chemical composition ratio suitable for spinning from basalt roving, and acidity coefficients are also considered suitable for spinning to continuous fibers.

(2) For inorganic fiber spinning, all non-determination must be achieved without any crystalline parts. It was achieved through rapid cooling by water in order to achieve the non-determination

of all the raw materials. The XRD confirmed the presence of the static and non-static materials and produced the non-static materials that did not show the peak.

(3) During furnace cooling, the crystalline parts accounted for almost all of the parts, so they continued to break off without smooth spinning. When cooled through air, the surface met with the air and cooled rapidly, but the inside gradually cooled down to become crystallized, resulting in short circuits of the fibers in the crystalline parts of the spinning, making it difficult to release them.

(4) The TG/DTA analysis shows that the melting point and glazing temperature do not show a constant relationship with the mixing ratio of slag, and the sample with the highest melting point shall have a minimum capacity of 1,320°C. Also, the temperature that can be spun through each glass transfer temperature can be expected to be more than 757°C. The temperature at which the viscosity of the melt becomes a viscosity suitable for spinning is not directly known, but the temperature at which the viscosity gradually decreases can be inferred through the temperature of the glass transfer temperature.

4.3 Spinning Viscosity

Viscosity of slags is a complex function of the slag composition, temperature and partial pressure of the oxygen in the system. Slags commonly contain four or more main chemical components that have a complex effect on viscosity. The determination of viscosities of the industrial slags is therefore a complicated task due to large number of variables and cannot be solved by means of experimental measurements only. The difficulty and high cost of measuring the viscosity of slags has led to the development of a number of viscosity models. These models can be used to predict the trends in viscosity as a function of the key variables, and so assist in the selection of process conditions and the optimization of the performance of slag fiber.[3, 7]

4.3.1 Urbain Model

The Urbain formalism is one of the most widely used slag viscosity models. The model is based on CaO–Al_2O_3–SiO_2 system and in the model the slag constituents are classified into three categories: glass formers (XG), modifiers (XM) and amphoterics (XA). The model assumes Weymann–Frenkel relation.[11]

$$\eta(p) = A\exp[1000B/T]; -lnA = mB + n \qquad (4.1)$$

where A and B are compositionally dependent parameters whereas m and n are empirical parameters. Urbain found that A and B were linked through the following equation: $-lnA = 0.29B + 11.57$. The parameters are shown in Tables 4.6 and 4.7.

The parameter B can be expressed by third-order polynomical Equation (4.2) whereas B_0, B_1, B_2 and B_3 can be obtained by Equations (4.3) and (4.4).

$$B = B_0 + B_1 X_{G1} + B_2 X_{G2} + B_3 X_{G3} \qquad (4.2)$$

$$B_i = a_i + b_i\alpha_1 + c_i\alpha_2 \qquad (4.3)$$

$$\alpha = X_M/(X_M + X_A) \qquad (4.4)$$

Table 4.6 Equations for B-parameters in Urbain model

B_0	$13.8 + 39.9355\alpha - 44.049\alpha2$
B_1	$30.481 - 117.1505\alpha + 139.9978\alpha2$
B_2	$-40.9429 + 234.0486\alpha - 300.04\alpha2$
B_3	$60.7619 - 153.9276\alpha + 211.1616\alpha2$

Table 4.7 Parameters a_i, b_i and c_i for MgO, CaO and MnO

	a_i	b_i			c_i		
i	All	Mg	Ca	Mn	Mg	Ca	Mn
0	13.2	15.9	41.5	20.0	−18.6	−45.0	−25.6
1	30.5	−54.1	−117.2	26	33.0	130.0	−56.0
2	−40.4	138	232.1	−110.3	−112.0	−298.6	186.2
3	60.8	−99.8	−1156.4	64.3	97.6	213.6	−104.6

where subscript i can be 0, 1, 2 or 3 and a, b and c are given constants for each case.

B_0, B_1, B_2 and B_3 can be calculated from the equations listed in Table 4.6. These parameters are then introduced into Equation (4.2) to calculate B. Parameter A can be calculated by Equation (4.2) and the viscosity of the slag can then be determined by using Equation (4.1). Urbain modified the model later to calculate separate B-values for different individual modifiers CaO, MgO and MnO and then the mean B is calculated. Table 4.7 shows the parameters a_i, b_i and c_i of Equation (4.3) for three cations. It is assumed that Fe_2O_3 and Cr_2O_3 behave both as network breakers and as amphoterics, where f is the fraction ba having network modifiers and a value $f = 0.6$ is assumed. The global B-value is given by;

$$B_{\text{global}} = (X_{CaO}B_{CaO} + X_{MgO}B_{MgO} + X_{MnO}B_{MnO})/$$
$$(X_{CaO} + X_{MgO} + X_{MnO}) \qquad (4.5)$$

4.3.2 *Prediction of Viscosity*

The viscosity of the molten solids consisting of 40, 50 and 60% of basalts and fly ash, respectively, was calculated to predict the critical viscosity temperature (Tcv) of each raw material and the viscosity most suitable for spinning. In general, the ideal spinning viscosity for smooth melt spinning of inorganic fibers, such as glass fibers and basalt fibers, is known as 102.8–103 poise. Slag fiber is also a fiber whose main ingredient is SiO_2, and it is estimated that the high viscosity will be smooth spinning at $10^{2.8}$–10^3 poise. To check the temperature of melt spinning, high temperature viscosity was predicted and the temperature was checked through the Urbain model.[12]

The material in which SiO_2 is the principal component represents a relatively high viscosity value because the structural unit forms a continuously connected mesh structure, and the higher the connectivity of these structures, the higher the viscosity. However, since CaO and MgO act as modifiers, the internal chemical composition ratio has a significant effect on viscosity.[2]

The Tcv of the combined slag can be predicted from the viscosity measurement results. Tcv refers to the temperature at which the viscous flow characteristics of slag change from Newtonian flow to non-Newtonian flow. At this temperature, the slag on the solid phase is first formed as a liquid slag, and above this temperature, the slag represents the plastic fluid flow characteristics. The melting viscosity of the blended slag depends on the content of SiO_2. The viscosity was high because SiO_2 content of slag with a content of 60 fly ash was the highest, and the viscosity of slag with 40% of fly ash with low SiO_2 content was the lowest. The Tcv of slag with a content of 40% fly ash was formed at a temperature lower than 800°C, and the Tcv of basalt gemstones was about 850°C, the slag with a content of 50% fly ash was formed around 880°C, and the slag with a content of 60% fly ash was formed at near about 900°C, showing viscous flow characteristics. The graph for this viscosity is shown in Figs. 4.7–4.9, where each viscosity difference is very high at 800°C. However, it can be inferred that the nearly similar viscosity at around 1,400°C indicates full melting at temperatures above about 1,400°C. Figures 4.7– 4.10 show the viscosities of samples.

Table 4.8 shows the fiberization temperature to spinning conditions. Temperatures with logarithmic viscosity values of 2.8–3 are referred to as forming temperatures in gas and are considered fibrous

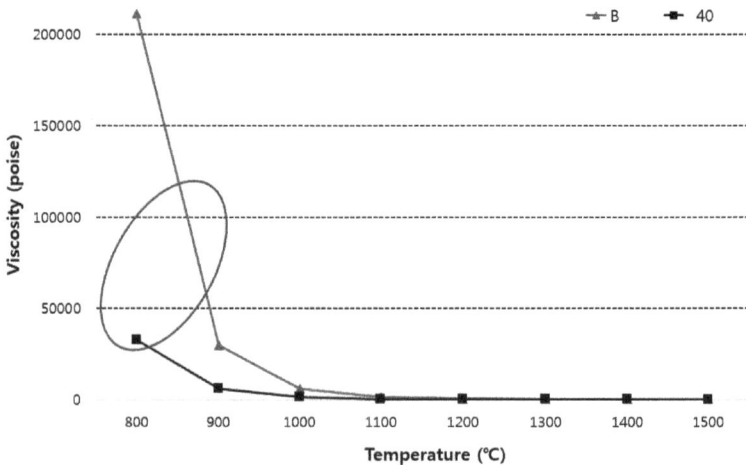

Fig. 4.7 Calculated viscosities of basalt and 40% fly ash.

Table 4.8 Fiberization temperature

Log 10 (Value)	3	2.8
Basalt	1,117.82° C	1,165.95° C
Slag (40:30:30)	1,025.11° C	1,062.58° C
Slag (50:25:25)	1,189.16° C	1,227.83° C
Slag (60:20:20)	1,258.50° C	1,297.32° C

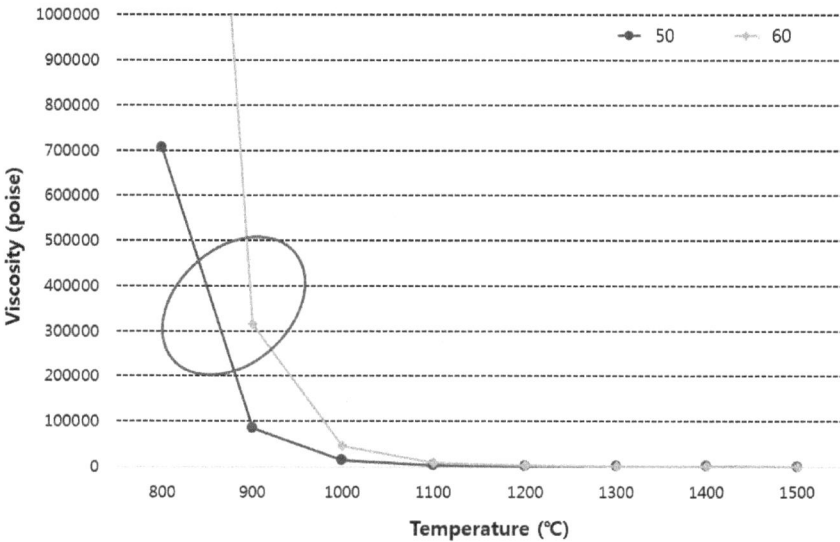

Fig. 4.8 Calculated viscosities of fly ash at 50% and 60%.

temperature in textile manufacturing. As shown in Table 4.8, the fiberization temperature of the basalts was about 1,127–1,165° C, the slag with 40 fly ash was about 1,125–1,062° C, the slag with 50 fly ash was about 1,189–1,227° C, and the slag with 60 fly ash was about 1,258–1,297° C, respectively.

4.3.3 *Conclusion*

The temperature suitable for each Tcv and spinning was derived through the high-temperature viscosity. The Tcv of the basalt was formed at about 850°C and the fibrous temperature was formed at

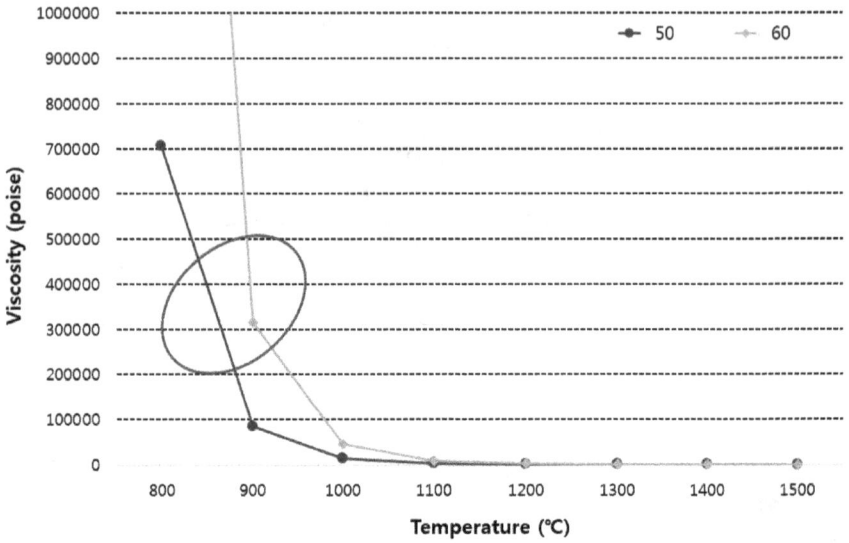

Fig. 4.9   Calculated viscosities of all samples.

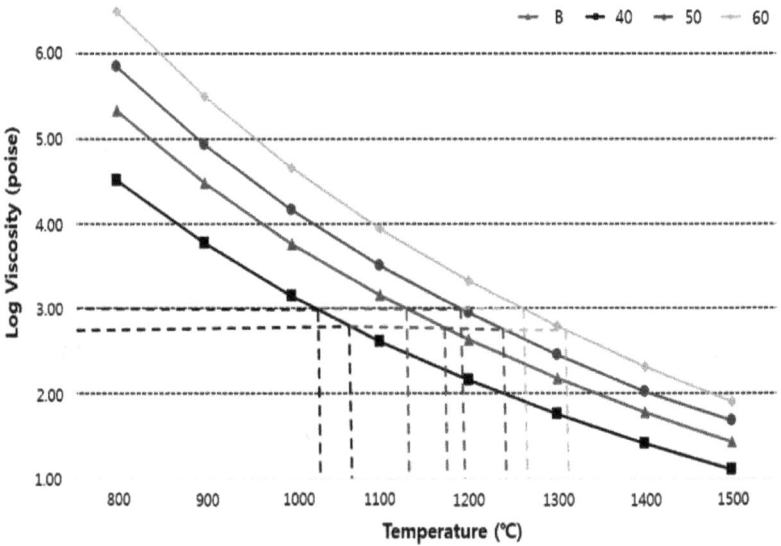

Fig. 4.10   Calculated viscosities of samples.

about 1,165–1,127°C. The Tcv of slag with fly ash of 40 was formed at a temperature lower than 800°C and the fibrous temperature was formed at 1,062–1,125°C. The Tcv of slag with a fly ash of 50 was formed at about 880°C and the fibrous temperature was formed at 1,189–1,227°C, and slag with a fly ash of 60 was formed at about 900°C and 1,258–1,297°C, respectively.

4.4 Spinning

While the melt spinning is simple, eco-friendly in manufacturing and its interest in application expansion is greatly increasing, its lower precision than carbon fiber spinning is a disadvantage. In the case of basalt fibers, most similar to slag fibers, research on the manufacturing process is insufficient and more research is needed. The fibers obtained from the melt spinning process in an ultra-high temperature environment have different process characteristics than other fibers. The fusion escapes the bushing and cools down at the same time as exhalation to become fibrous. The accumulation and production technology of data based on empirical and theoretical production technologies can lead to the development of the production technology of fibers spun in various ultra-high-temperature environments.

4.4.1 *Spinning Machine*

Generally, a spinning system needs an industrial furnace with a high heat capacity as the base material of the slag is melted at a temperature above 1,300°C. It also requires precise controls of heating rate, temperature and diameter of the fiber. In this study, a feasible lab scale spinning system, which meets above requirements, has been designed. It consists of three parts: furnace, bushing and winding part.

The spinning system has been designed considering the fact that the compounded slag is mainly composited with SiO_2 with high amounts of other compounds such as of Al_2O_3. The furnace part can raise the temperature up to a maximum of 1,600°C, and the refractory of the furnace has been crafted with heat resisting and insulating materials that are frictionless and thermostable at high

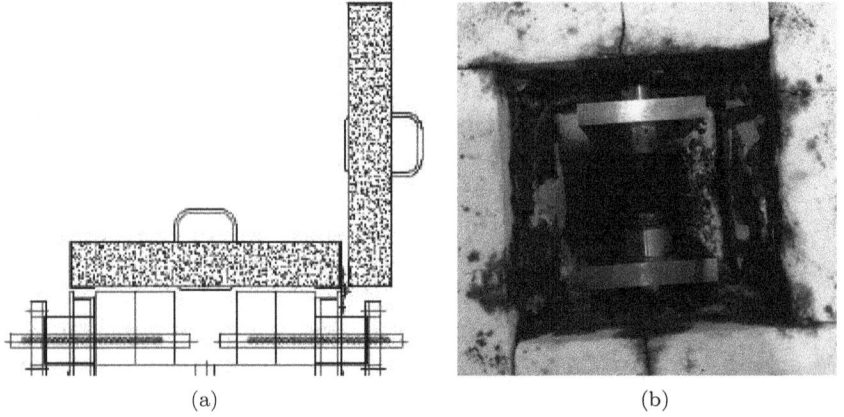

(a)   (b)

Fig. 4.11   Furnace part.

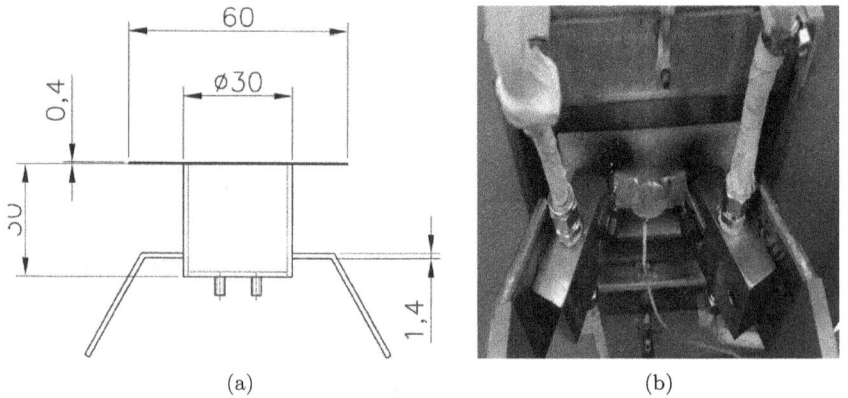

(a)   (b)

Fig. 4.12   Bushing part.

temperature. Figure 4.11(a) presents the components of the furnace part, and (b) depicts the actual furnace part in lab scale.

As this is the most core construction of this spinning system, the bushing is made with an alloy of Pt–Rh. The design includes two holes based on the lab scale. Also, this part is designed thermostatically, in order to prevent hardening and clogging of the bushing entrance. Figure 4.12(a) shows the design specification of bushing and (b) the actual bushing part of this design.

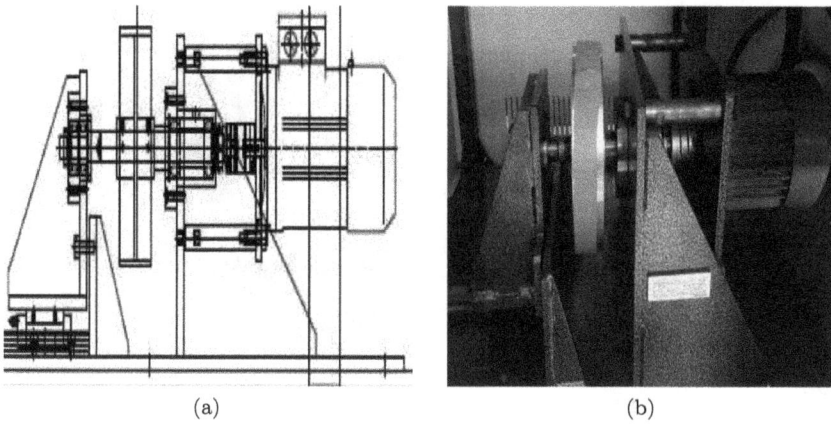

(a)   (b)

Fig. 4.13   Winding part.

Winding system consists of two bobbins and a guide, and has a speed control capacity of 0–6,000 rpm. This part has been designed to wind a superfine fiber of 6–24 μm diameter without breaking it. Also, when a fixed amount of fiber is winded, the bobbin will shift automatically, allowing further winding. Figure 4.13(a) shows the design specification of winder and (b) the actual winding part of this design.

4.4.2   *Spinning Process*

(a) Particulateization of raw materials:
    After grinding the mixed slag first through a rock shredder, the raw material is ball milled so that the overall homogenization of the composition component can be achieved through a slip of 400–800 μm. Migrated basalt and mixed slag are shown in Fig. 4.14.

(b) Melting temperature control and primary melting:
    Heat it at 1400°C for 2 h using a customized electric furnace. These solvents are homogenized through this process and release foreign substances in various gaseous states. After that, the crucible is taken out of the furnace and cooled sharply using

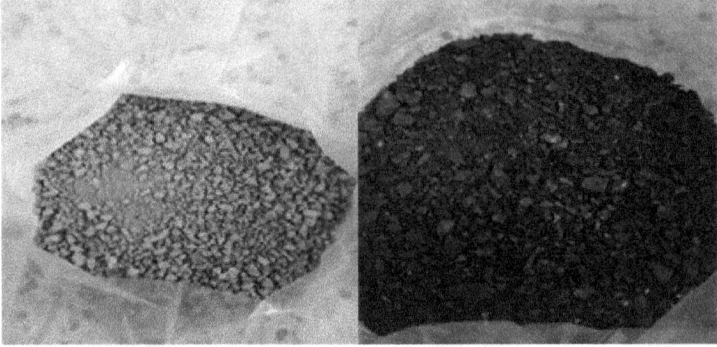

Fig. 4.14   Atomization of raw materials.

water to manufacture the raw material of glass. Use ball mill to atomize the manufactured material.

(c) Secondary melting and maintaining temperature:

After the first melt, put the raw material that has been particulateized into the bushing and maintain it at a fibrous temperature for a certain period based on viscosity results. Once the bushing reaches the fibrous temperature, and a floating layer appears, similar to the formwork, separated from the weld pool, removed through physical force and, over time, it becomes the same form as Fig. 4.15. The spinning was carried out after 2 h of retention after the fiber temperature was reached.

(d) Spinning:

It is controlled to maintain at an appropriate fiberization temperature. After pulling the weld pool on the tip nozzle of the bushing, adjust the winding speed to control and spin the diameter. Figure 4.16 shows spun slag fibers.

The amount of weld through the nozzle hole of the bushing tip can be determined by Poissieule's equation.

$$F = \pi \rho g h / 8\eta L \tag{4.6}$$

F: Amount of melt through nozzle of bushing per unit hour

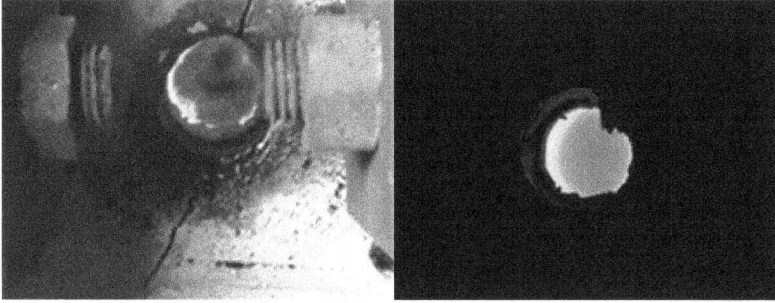

Fig. 4.15 Melting of raw materials.

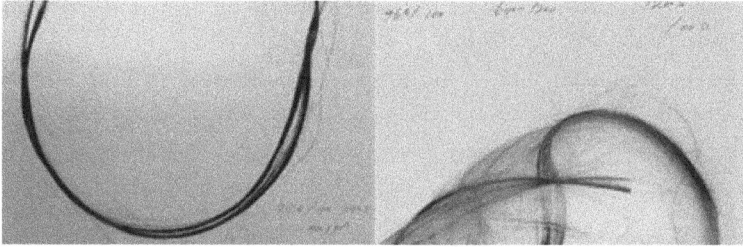

Fig. 4.16 Fiberization.

ρ: Specific gravity of melt
g: Gravitational acceleration
h: Height from bushing to winding machine
η: Viscosity of melt
L: Length of bushing tip

In the aforementioned expression, the viscosity of the melt and the length of the bushing tip determine the amount of melt through the nozzle of the bushing per unit hour. The specific gravity of melt, gravity acceleration, height from bushing to winding machine, the length of the bushing tip are all fixed and the viscosity determines the amount of weld through the nozzle of the bushing per unit hour. The speed of winding also adds to the acceleration of gravity, which changes the amount of melt flowing out of the nozzle. The viscosity can be changed by the temperature of the melt, and the acceleration of the melt may vary by the speed of winding, so the amount of the melt changes by the temperature of the melt, i.e., the temperature of the fiber and the rate of winding, so the diameter of the fiber is controlled. Therefore, the diameter of the fiber is determined by the temperature of the spinning and the winding speed of the winding machine, so continuous microfiber spinning can be achieved only when the spinning temperature and the winding speed are strictly controlled.[5, 16]

The bushing temperature was spun by adjusting the speed with each fiberization temperature. Table 4.9 shows the change of slag fiber diameter according to spinning conditions. The result values were written down for more than one minute in a row, otherwise it was considered to be a failure. The spinning test showed that for a content of 50% fly ash, the spinning temperature was at least 1,220°C up to 1,224°C and averaged 1,221°C. It was also reliably spun at an average winding speed of 16 m/sec and fibers with

Table 4.9 Slag fiber diameter change according to spinning conditions

Fly ash:Furnace Fe–Ni	Spinning temp. (°)	Winding speed (m/sec)	Diameter (μm)
40:25:25	1,150	9.93	32.63
	1,180	Fail	Fail
	1,155	10.01	25.46
	1,167	Fail	Fail
	1,152	8.24	28.32
	1,223	16.22	20.34
	1,220	15.84	24.45
50:25:25	1,221	15.48	25.01
	1,224	16.04	22.34
	1,220	16.43	18.54
	1,245	6.49	32.20
	1,250	7.04	31.73
60:25:25	1,245	6.58	32.45
	1,247	6.50	33.43
	1,247	6.49	32.51

an average diameter of 22.14 μm were spun. If the content of the fly ash is 60, fibers with an average diameter of 32.46 mm were spun under the conditions of 1,260°C and 6.62 m/sec. If the fly ash proportion is 40, it was not spun due to high viscosity at the fibrous temperature, and the spinning temperature was re-discovered by raising the temperature to lower the viscosity. As a result of the spinning, two out of five failed and only three succeeded. Fiber with an average diameter of 28.80 mm was spun under conditions of 9.39 m/sec on average at 1160°C. If the content of the fly ash is 50 or 60, the predicted log values of the fibrous temperature spin at a temperature of 2.8–3, but in the case of 40, they are not spinning at a temperature of 1,025–1,062°C and are unstable at a higher temperature. The diameter of the spinning fiber was measured using SEM and shown in Fig. 4.17.

Fig. 4.17 Diameter of slag fiber by FESEM.

4.4.3 Conclusion

In this section, we predicted the Tcv, the appropriate spinning titration temperature and conducted melting spinning of the mixed slag to identify the spinning process and characteristics.

(1) High-temperature viscosity was the highest in the slag with a content of 60 fly ash, which has the highest content of SiO_2, and high-temperature viscosity was shown according to the order of SiO_2 content. Since SiO_2 acts as a continuous connected mesh structure and CaO and MgO act as modifiers, the main component is SiO_2, whose effect is believed to be the greatest.

(2) To fiberize slag, slag fiber spinning divided into the melting, bushing and winding systems were designed and manufactured, and continuous microfiber spinning was possible through temperature control of bushing and adjustment of winding speed.

(3) The slag fiber spinning process is to particulateize the raw material and release the homogenization of the raw material and foreign substances in various gaseous states through primary melting. The first molten raw material is glass-stained through rapid cooling using water. After 2 h of melting, the raw material that was glassed, it was maintained at the temperature of fiberization for 2 h, and then the winding speed was adjusted and the fiber was released.

(4) The average spinning temperature of 1,221°C and the average winding rate of 16 m/sec. were spun in the case of slag with 50% of fly ash. The diameter was 22.14 μm on average, the most stable of the three cases, and the most flexible was possible by controlling the winding speed. In the case of content of 60% fly ash, fibers with an average diameter of 32.46 mm were spun under the conditions of 1,260°C and 6.62 m/sec. And in the case of fly ash content being 40, it was not spun due to high viscosity at the fibrous temperature, and the spinning temperature was found again to lower the viscosity.

4.5 Slag Fiber Properties

Glass fiber, carbon fiber, aramid fiber and basalt fiber, which are currently used in the industries, have been developed and detailed process variables have been established through various fibrous process research. In addition, the database was accumulated through repeated research results such as evaluating the characteristics of element materials, testing material properties and environment, securing processes suitable for industrial application and securing characteristics. The reliability of each material was secured through the development of accumulated databases and new technologies. However, the fundamental problem with raw materials is difficult to solve. In the case of glass fibers, all B_2O_3 are imported from abroad to spin glass fibers, and in the case of basalt fibers, they are rich in reserves, but there is a limit to the use of limited natural resources as raw materials.

In the case of slag fiber, however, unlike glass fiber and basalt fiber that use steel slag as raw material, the cost of raw material is not consumed, so the price competitiveness is excellent and the more the slag fiber produced, the better the treatment of waste slag. It is a technology that can be symbiotic with the steel industry by reducing the cost of disposal of steel slag accordingly. In addition, a new method of recycling can be presented by combining various process by-products, unlike the recovery of metals or resources from existing single-creation waste resources.

In this section, basic physical data of slag fibers were accumulated and the characteristics of fibers were studied through tensile strength and environmental testing.

4.5.1 *Environment Properties*

(a) Absorption and specific gravity test: The proportion of the combined slag was 2.98 in surface dry saturation. Generally, glass fibers have a proportion of 2.5 and basalt fibers 2.5–3. Carbon fiber has a proportion of 1°, and slag fiber has a higher proportion than carbon fiber and has a similar proportion to glass and basalt fiber. The absorption rate of the mixed slag was 0.05°, indicating that the water absorption rate was less varied over 48 h.

(b) Chemical resistance test: Based on the mass reduction rate after 24 h of drug treatment, $BaCl_2$ of 37.5% had the highest mass reduction rate and the lowest mass reduction rate of 20% NaOH. All of them showed a mass reduction rate of less than 0.1%, and they are considered to have excellent resistance to drugs. Table 4.10 shows the pre-treatment mass, post-treatment mass and reduction rate.

(c) Wear test: The wear rate was 1.194% and the non-wear rate was 3.15×10^{-7} m^3/N, showing very good characteristics. The results for the wear test are shown in Table 4.11. In the case of mixed slag, the wear characteristics were excellent.

Table 4.10 Result of chemical resistance test

	Before weight (g)	After weight (g)	Reduction rate (%)
$BaCl_2$(37.5%)	50	49.955	0.09
HCl (0.25%)	50	49.98	0.04
HCl (22.37%)	50	49.99	0.02
NaOH (20%)	50	49.995	0.01
NaOH (50%)	50	49.985	0.03
H_2SO_4 (75%)	50	49.975	0.05
H_2SO_4 (98%)	50	49.98	0.04

Table 4.11 Result of wear test

Disc weight (g)	Before	15.5847
	After	15.3986
	Variation	−0.1861
Ball weight (g)	Before	8.3673
	After	8.3373
	Variation	−0.03
Mass change rate (%)		1.1941
Non-wear (m^3/N)		3.15×10^{-7}

4.5.2 *Tensile Strength Properties*

There are two experimental measurement methods in Short Fiber Tensile Strength: ASTM D 3822 and ASTM D 4018. ASTM D 3822 is a method of preparing every strand of specimen by separating spun fibers and then using paper as a frame. In this method, a tensile test is conducted at 1.0 mm/min in a small tensile tester. At least 50 data items are required because the data error is large. Therefore, the short fiber tensile strength was tested by ASTM D 4018 instead of ASTM D 3822. All the filaments were washed using the acetone in order to clean the fiber surface. Tensile strength was tested by ASTM D 4018 strand tensile test. This test is to fabricate specimens with fiber bundles. Orientation of the fiber is in one direction and the fiber content produced has a 40–50 wt%. Since the tensile strength of resin is very small compared to fiber, tensile strength was calculated by ignoring resin. Figure 4.18 shows a schematic representation of the ASTM D 3822 and ASTM D 4018.[4, 8, 10]

By comparing the fiber spinning and tensile strength of the three types of fly ash content, varying at 40, 50 and 60 wt.%, the most suitable component mix ratio for fibrosis was derived (Fig. 4.19). The tensile strength of the fiberized slag fiber was measured by the spinning experiment by viscosity prediction. Slag with a content of 40% fly ash failed two out of five emissions, and the average tensile strength of three spinnings was 1,122 MPa. Slag with a content of 50% fly ash was the most stable of the three, with an average tensile strength of 2,873 MPa, the highest tensile strength. Finally, the average tensile strength of slag with a content of 60 fly ash is 964.6 MPa. Table 4.12 shows maximum tensile load and tensile strength.

In the course of the spinning experiment, mixed slag with a content of 40 fly ash was spun at a higher temperature than the predicted viscosity, and it was also difficult to spin continuous fibers, especially when spinning was not produced. In the case of spinning fibers, the winding speed was slower than that of a specimen with a content of fly ash of 50 and the larger diameter fibers were spun. In the case of mixed slag with a content of 60 fly ash, it was spun at the predicted viscosity, but was cut off when the winding speed

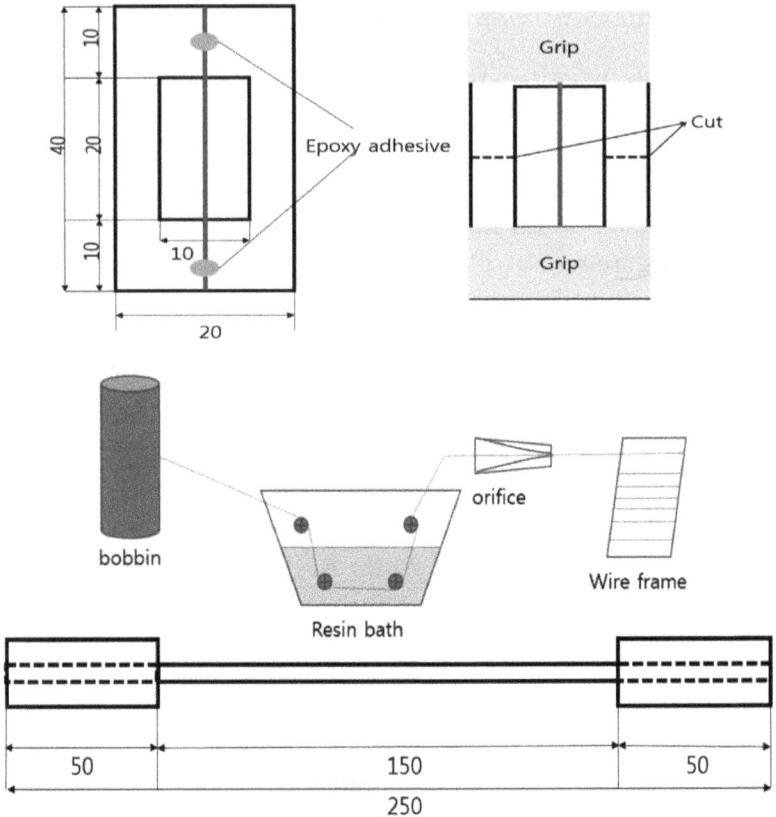

Fig. 4.18 Schematic representation of single fiber tensile test (ASTM D 3822 and 4018).

was increased, so the diameter of the fiber could not be controlled, and the spinning was possible without being able to deviate from the speed of about 6.5–7 m/sec. On the other hand, when mixed slag with a content of 50 fly ash was released, the log value known as the fibrous temperature could be released in a fluid manner at 1,189–1,227°C, a temperature between 2.8 and 3, enabling diameter control. The fiber was spun at the most stable winding speed, with an average speed of about 16 m/sec; the fastest of three possible windings and the most stable of all. Also, depending on the winding speed, the diameter of the fiber also varies.

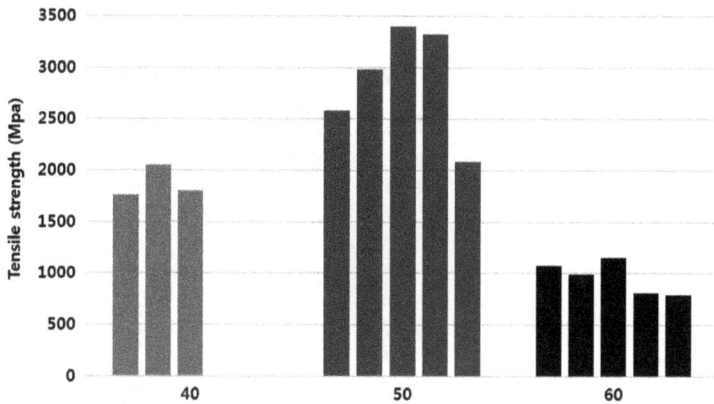

Fig. 4.19   Comparison of tensile strength with fly ash content.

Table 4.12   Maximum tensile load and tensile strength

Fly ash:Furnace: Fe-Ni Mixing ratio	Maximum tensile load (N)	Tensile strength (MPa)
40:30:30	252.91	1,760
	—	—
	249.58	2,050
	—	—
	258.66	1,800
50:25:25	247.16	2,580
	285.48	2,980
	325.72	3,400
	318.05	3,320
	199.74	2,085
60:20:20	144.85	1,080
	123.17	989
	176.58	1,152
	77.79	812
	128.66	790

The combination ratio best suited for fiber spinning of the three types of fly ash content varying by 40, 50 and 60% was studied and the most spinnable, with good tensile strength was shown when the ratio of fly ash:furnace slag:Fe–Ni slag was 50:25:25. Continuous

Table 4.13 Diameter of slag fiber on spinning conditions

	Spinning temp. (°C)	Winding speed (m/sec.)	Diameter (μm)
1	1,180	38.94	16.8
2	1,220	24.66	20.8
3	1,221	25.66	20.5
4	1,223	26.24	19.8
5	1,223	25.96	19.8
6	1,225	26.61	19.5
7	1,195	23.36	20.8
8	1,197	25.9	20.5
9	1,197	25.96	19.8
10	1,203	27.26	18.5
11	1,201	26.61	18.5
12	1,205	29.85	17.5
13	1,210	25.96	20.5
14	1,220	24.66	20.5
15	1,240	26.3	19.8
16	1,205	29.2	17.5
17	1,230	26.1	20.5
18	1,230	29.2	18.2
19	1,231	23.36	22.7
Min.	1,180	23.36	16.8
Max.	1,240	38.94	22.7
Average	1,213.47	26.94	19.61

fiber spinning was conducted with a mixed slag with a content of 50% fly ash, which is most suitable for continuous fiber spinning, and the characteristics of the fiber spinning process and tensile strength were recognized.

Based on the high-temperature viscosity prediction model of slag, the number of cases where fibrous temperature was determined continuously for more than one minute was recorded, and the diameter of the fiber was measured using an optical microscope.

The results of the spinning experiment were shown in Table 4.13. The spinning temperature was spun in a wider range than in the range of 1,189–1,227°C, a temperature in which the logarithmic value in the high-temperature viscosity prediction model was 2.8–3. Smooth spinning was carried out in all between 1,180°C and 1,240°C

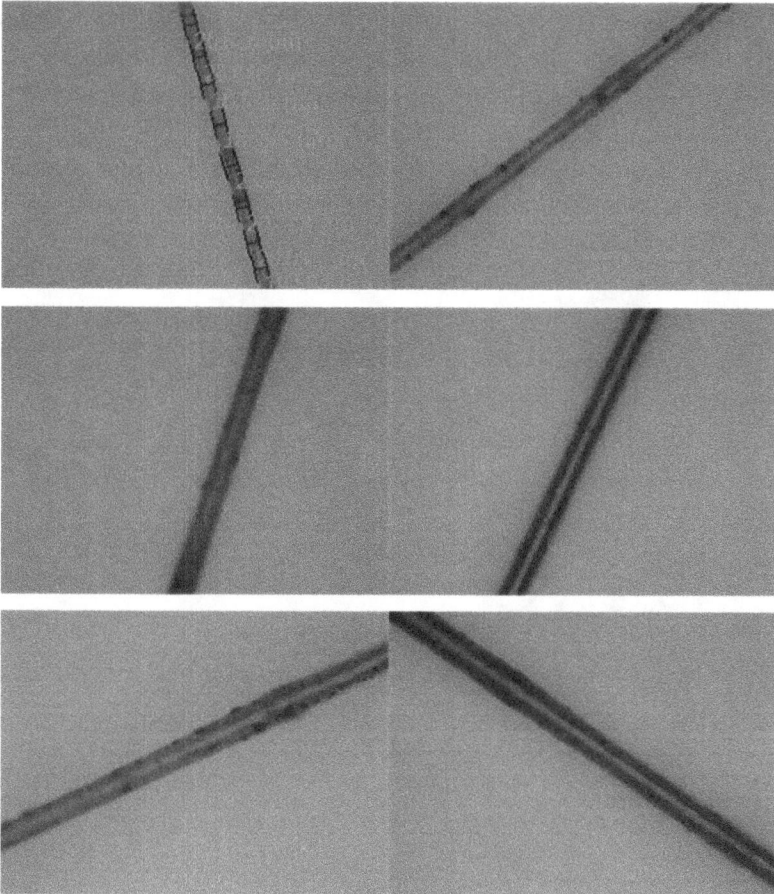

Fig. 4.20 Diameter of slag fiber by optical microscope.

and spun at an average of 1,213°C. The winding speed was from 23.36 to 38.94 m/sec.

The diameter of the fiber is measured by using an optical microscope to measure 6–8 parts after the fiber spinning, and the average value is taken as the diameter. The optical microscope image of the fiber is shown in Fig. 4.20. The diameter of the fiber averaged 19.61 mm and varied according to the winding speed. Winding speed and the diameter of the fiber showed a tendency of inverse proportion, i.e., the faster the melt exits from the tip of the bushing, the smaller the diameter of the fiber. If the winding speed is similar, the diameter is smaller if the spinning temperature is high. In other words, the diameter of the fiber is affected by the winding speed and the temperature of the bushing, which is the fibrous temperature, which has a greater influence on the winding speed and a smaller effect on the temperature than the effect on the temperature of the fiber. Effect on diameter of fibers is shown in Figs. 4.21 and 4.22.

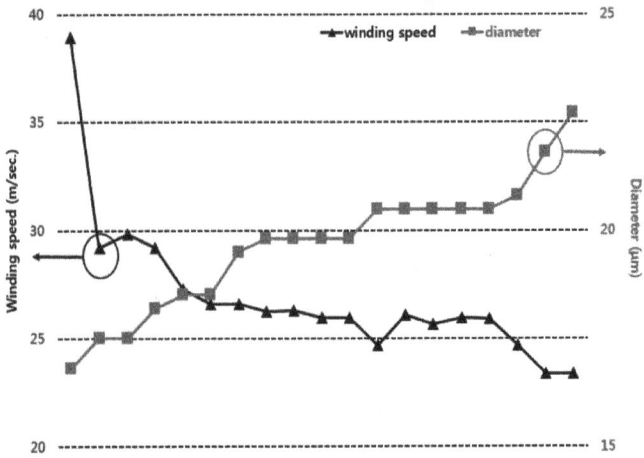

Fig. 4.21   Diameter of slag fiber on winding speed.

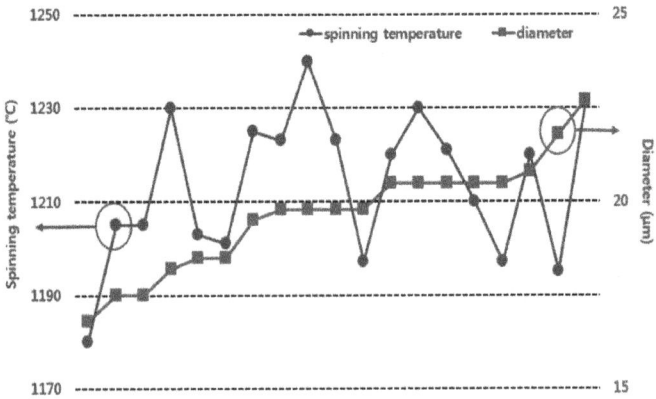

Fig. 4.22   Diameter of slag fiber on spinning temperature.

The winding speed and diameter of the fiber were almost inversely proportional. On the other hand, the temperature of the melt and the diameter of the fiber did not show a constant tendency.

The tensile strength of slag fiber with a content of 50% fly ash was 1,960 MPa, a minimum value, and 4,176 MPa, a maximum value. In general, the difference in tensile strength of a single fiber is caused by defects such as air bubbles in the fiber or by variables of fine crack and stress in the cooling process. In addition, the fiber spinning device used in the research is a lab scale experimental device that can emit only a small amount of spinning during the spinning experiment. Even if the composition ratio is the same, the conditions of the fibers vary depending on the conditions of the experiment. The average tensile strength is about 3,000 MPa, and in Fig. 4.23, data are presented.

The tensile strength of single fiber was compared with E-glass fiber, which is the main component of SiO_2 and is widely used in the industry. E-glass and basalt fibers were heat treated at 350–400°C within the electric furnace to remove the sizing agent used in the fiber. In the case of slag fibers, the raw materials were released without separate sizing treatment, so each of the 10 specimens were produced without sizing removal, and the average value is shown

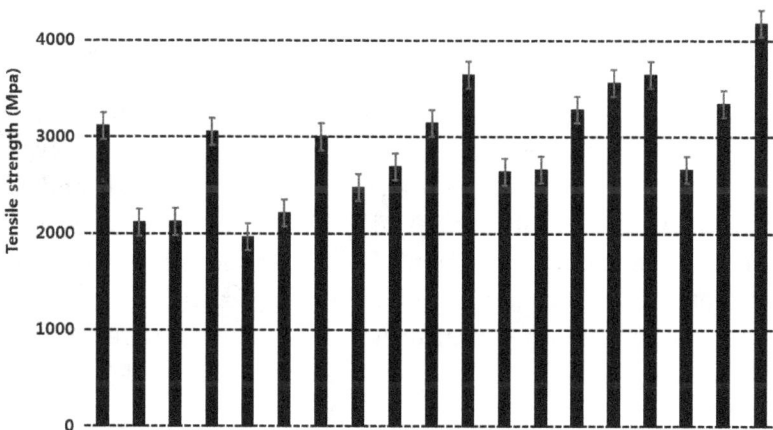

Fig. 4.23   Tensile strength (fly ash 50 wt%).

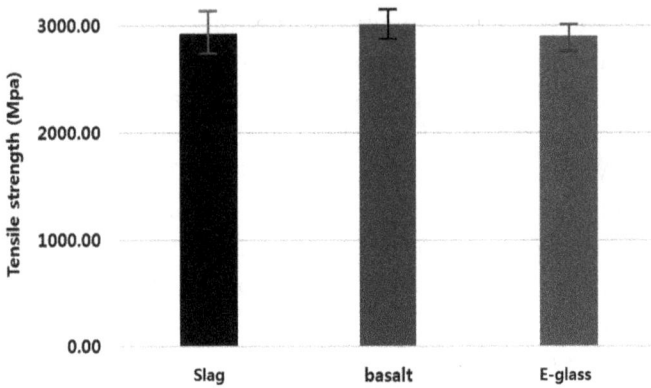

Fig. 4.24 Tensile strength of slag fiber, basalt and glass fiber.

in Fig. 4.24. The average tensile strength value of single fiber was approximated to 3,000 MPa for slag, basalt and glass fiber. Single-fiber tensile strength is thought to have individual deviation, but slag fiber also has enough strength to be used in the industry. In the case of slag fibers, the difference in tensile strength of each specimen was large compared to other fibers. In the case of basalt and glass fibers applied to and sold in the industry, large amounts of fiber can be emitted in a single process because they can be mass produced, so the error is thought to be smaller than slag fibers.

4.5.3 *Acid and Alkalinity Properties of Slag Fiber*

The effects were examined by treating them in NaOH and HCl solutions represented by basic and acidic solutions, respectively. The fiber was cleaned by using acetone to clean the surface and then treated in 10 wt% NaOH and 10 wt% HCl. The treatment time was 1–8 h, and the experimental solution was kept at room temperature throughout the entire experimental process. Tensile specimens were produced after solution treatment, the filaments were immersed in distilled water for 24 h, and dried at 100°C for 4 h using an oven.[1, 6, 7, 17]

Figure 4.25 shows the behavior of the mass loss ratio over the bedding time in HCl. The three types of fibers tend to be similar, all

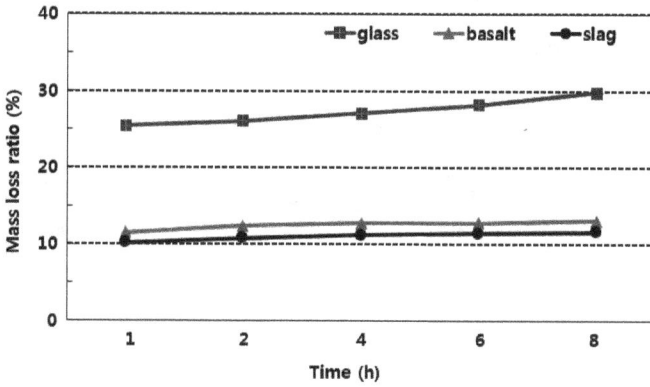

Fig. 4.25   Mass loss ratio on immersion time (HCl solution).

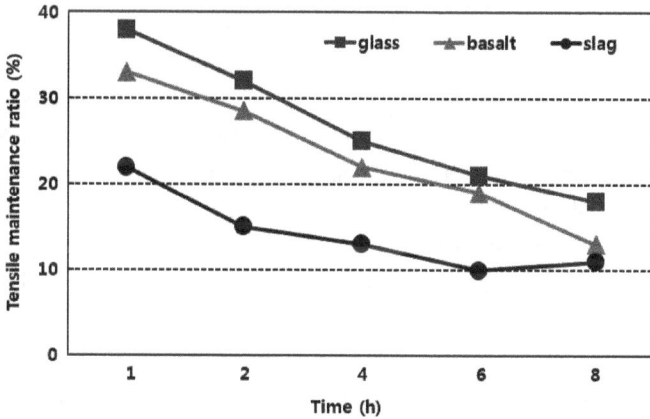

Fig. 4.26   Tensile maintenance ratio on immersion time (HCl solution).

of which lose mass sharply over an hour. The basalt and slag fibers showed nearly similar mass reduction rates after an hour, but in the case of glass fibers, the mass loss rate continued to increase up to 8 h. Figure 4.26 shows tensile behavior after HCl treatment. The tensile strength of the three fibers decreased sharply up to an hour. After that, tensile strength has shown behavior similar to the rate of loss of mass gradually decreasing. Based on the mass loss rate and tensile strength behavior, basalt and slag fibers showed stronger acid resistance than glass fibers.

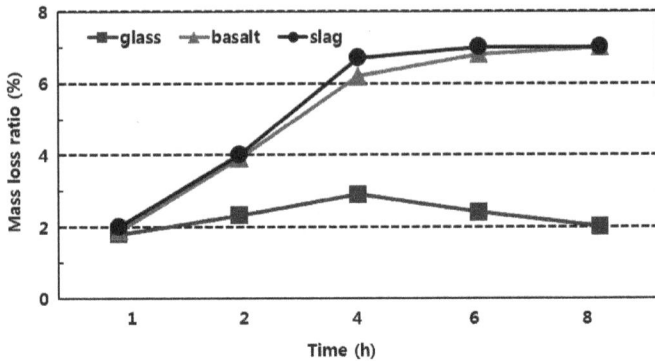

Fig. 4.27 Mass loss ratio on immersion time (NaOH solution).

Figure 4.27 shows the behavior of the mass loss ratio over the bedding time in NaOH. Basalt and slag fibers showed similar tendencies. Within 4 h the mass reduction rate increased rapidly and then showed a constant mass reduction rate. For glass fibers, the reduction rate tended to be smaller than basalt and slag fibers. Then, after 4 h, the mass reduction rate started to decrease. Figure 4.28 shows tensile behavior after NaOH treatment. Tensile strength showed a similar tendency for all three types of fibers. The tensile strength decreased sharply up to 1 h, and decreased steadily thereafter. Analysis of mass loss rate and tensile behavior showed that the alkaline resistance of basalt and slag fibers was lower than that of glass fibers, as opposed to HCl treatment.

Figure 4.29 shows variation of surface after burying slag fibers in HCl and NaOH for 8 h. Before HCl and NaOH treatment, the surface has a generally smooth surface, but corrosion progresses on the surface after treatment and is ionized from the fiber which is seen attached to the surface. This seems to have led to deterioration of the mass and tensile strength of the fiber, and the amount of residue on the surface of the fiber tends to be more attached to NaOH than to HCl. It can be seen that the reduction in mass at NaOH and the decrease in tensile strength are smaller than when the HCl solution is used for submerging. SEM analysis also shows that the amount of residual material on the surface of the fiber is higher, so slag

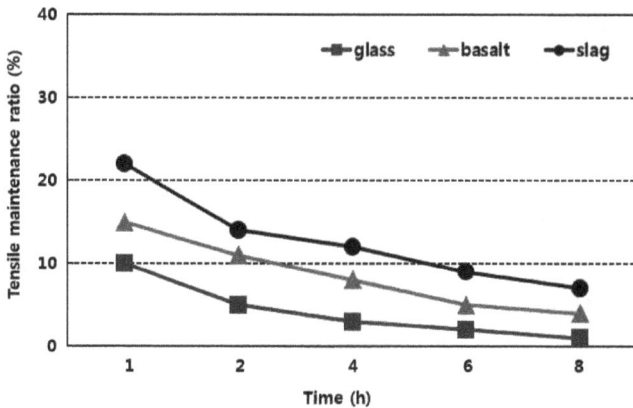

Fig. 4.28 Tensile maintenance ratio on immersion time (NaOH solution).

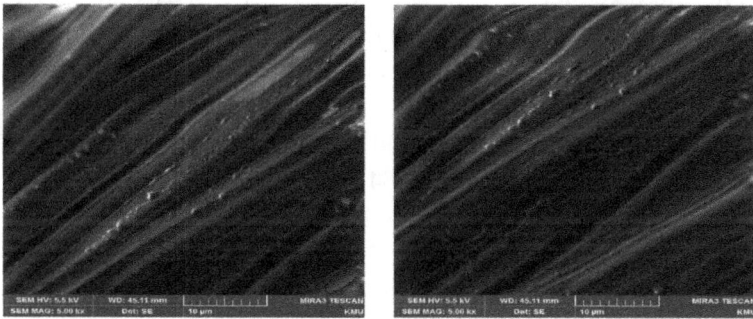

(a) Before

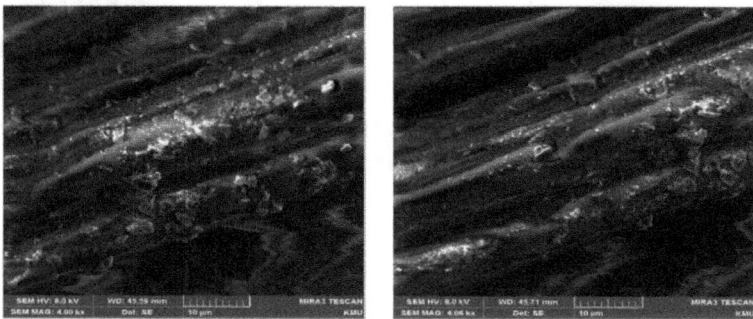

(b) HCl after   (c) NaOH after

Fig. 4.29 SEM images of slags fiber (a) before and after treatment in (b) HCl solution and (c) NaOH.

fiber is considered to be superior in chemical resistance in the NaOH environment to the HCl environment.

4.5.4 Conclusion

In this section, the characteristics of slag fibers through melt spinning were explored through absorption rate, specific gravity, resistance, wear test, environmental test and tensile strength test for application to the industry.

(1) The proportion of blended slag is 2.98 which is similar to glass, and basalt fiber used in the industry, which has a proportion of 2.5–3. The water absorption rate was 0.05% over 48 h, indicating very low absorption rate.

(2) The mass reduction rate of $BaCl_2$, HCl, NaOH and H_2SO_4 all showed a reduction rate of 0.1% or less, indicating excellent wear resistance with a very high resistance and a 1.194% wear rate.

(3) Fiber containing 40 fly ash showed an average tensile strength of 1,122 MPa, fiber containing 50 fly ash was 2,873 MPa, and fiber containing 60 fly ash was 964.6 MPa, showing highest tensile strength and excellent spinnable properties.

(4) Mixed slag with a content of 50 fly ash was spun at a spinning temperature of 1,189–1,227°C, winding speed of 23.36–38.94 m/sec and average diameter of 19.61 mm.

(5) The diameter of the fiber showed a semi-proportional tendency to get smaller as the winding speed was faster, and the higher the spinning temperature, the smaller the diameter, but there was no constant proportional relationship, and the diameter tended to get smaller when the winding speed was the same. The diameter of the fiber had a greater effect on the winding speed than the spinning temperature.

(6) The tensile strength of slag fiber with a content of 50 fly ash has the minimum value of 1,960 MPa, the maximum value of 4,176 MPa and has an average strength of 30,000 MPa. The difference between the minimum and maximum values is believed to have been caused by defects such as blisters in the fiber or by variables of fine crack and stress in the cooling process.

(7) Comparing tensile strength with glass, basalt fiber, which is widely used in industry, had similar strength values of 3,000 MPa. For all three types of fibers, the variation in tensile strength was large, but the strength deviation of slag fibers was larger. In the case of basalt and glass fibers, they can be mass produced, so they are believed to have a smaller deviation than slag fibers.

(8) In HCl environment, the mass loss rate of glass fibers was greater than that of basalt and slag fibers. The reduction rate of tensile strength was 80% for glass fibers and 65% for basalt and slag fibers. The basalt and slag fibers showed strong acid resistance compared to glass fibers.

(9) In the NaOH environment, the mass loss ratio increased sharply in basalt and slag fibers up to 4 h and then showed a constant rate of reduction in the quantity. In the case of glass fibers, the reduction rate tended to be smaller than other fibers, and the reduction rate of tensile strength showed a sharp decrease rate up to 1 h, followed by a constant decrease rate. In contrast to the HCl treatment, the alkaline resistance of basalt and slag fibers was less than that of glass fibers.

(10) Before HCl and NaOH treatment, the surface is generally smooth, but corrosion progresses on the surface after treatment and ionizes from the fiber and gets attached to the surface. This seems to have led to deterioration of the mass and tensile strength of the fiber, and the amount of residue on the surface of the fiber tends to be more attached to NaOH than to HCl.

(11) It can be seen that the reduction in mass at NaOH and the decrease in tensile strength are smaller than when the HCl solution is used for submerging. SEM analysis also shows that the amount of residue on the surface of the fiber is higher, so slag fibers are thought to be superior in their chemical resistance in the NaOH environment to the HCl environment.

4.6 Slag Fiber Absorption Behavior

Polymer composites are a promising material for structures and are widely utilized in various industries even under extreme conditions like an acid-base, seawater, corrosion environment. With a high durability, it has a complex failure behavior and design diversity of constituent materials. Recently, there have been attempts to transition from a paradigm that relies on high specifications and quality in terms of economic and environmental efficiency. Despite various studies on biomaterials, there are still conflicting aspects such as lifespan and safety of the materials, and pre- and post-environmental hazards when disposing of materials depending on the type of fiber reinforcement in the polymer matrix fiber reinforced plastic (FRP) composite.[9, 13] In this aspect, basalt fiber based on natural rock and slag fiber produced through refining and fiberization from slag, a by-product of the steelmaking process, exhibit a similar chemical composition and also have a relatively superior comparative advantage to carbon fiber and glass fiber with high compatibility. However, the slag fiber has not reached the commercialization stage. And this slag fiber, compared with basalt fiber, still lacks practical research such as physical performance, environmental resistance, chemical stability and functionality for structure, and also requires a feasibility analysis on alternative materials for the existing fiber reinforcement. Moreover, the similarity of the physical properties of basalt fiber and slag fiber is an important factor in securing the potential impact of slag fiber, and it will contribute to establishing a practical research infrastructure through material-based characteristic analysis under various loads and environmental conditions.

Therefore, in this study, to determine the effect of chemical composition on physical properties, customized slag fiber was used to analyze the similarity with basalt fiber on the environmental degradation characteristics. And it was carried out through water uptake behavior under fresh water/seawater environments. Furthermore, the mechanical properties of the basalt fiber and slag fiber were compared under tensile load and their failure causes were investigated.

4.6.1 *Experimental Absorption Behavior*

Two types of fiber reinforcements used for epoxy impregnated filament composites were basalt fiber of GM Composite Co., and customized slag fiber, which had a composition mixed with fly ash:furnance:Fe–Ni slag at 50:25:25 each and spinning at 16 m/sec. speed at an average temperature of 1,221°C through fiberization process. The polymer matrix was mixed with KFR-120V epoxy and KFH-141 hardener supplied from Kukdo Chemical Co. Ltd, at a ratio of 72:28. The water uptake behavior of these BFRP (basalt FRP) and SFRP (slag FRP) under fresh water/seawater absorption environments was measured through Equation (4.7) of Fickian Diffusion.

$$M = \frac{W_m - W_d}{W_d} \times 100 \qquad (4.7)$$

M: Absorption rate (%)
W_d: Initial specimen weight (g)
W_m: Specimen weight after immersion (g)

The tensile test was conducted and the removal-oriented tensile specimen with a fiber content was about 40–50 wt%. These tests were done by ASTM D 5229 and ASTM D 4018, respectively. The surface of the composites was observed through SEM to investigate the effect on damage behavior during fresh water/seawater immersion.

4.6.2 Moisture Absorption Behavior of BFRP and SFRP[18]

As shown in Fig. 4.30(a), BFRP and SFRP have an increased absorption rate in both fresh water and seawater, and maintained equilibrium by reaching water saturation state after 350 h. BFRP showed relatively high absorption rate in fresh water and seawater, whereas SFRP exhibited excellent hygroscopicity in fresh water. In terms of moisture absorption resistance, SFRP was 46.75% and 22.2% higher than BFRP in fresh water and seawater, respectively. SW-BFRP and SW-SFRP tended to be 14.9% and 67.2% lower than W-BFRP and W-SFRP. As a result, BFRP and SFRP were observed to be vulnerable in seawater. This is considered to be due to the relatively lower diffusion activity energy of seawater than that of fresh water at the same temperature. This resulted in an accelerated phenomenon

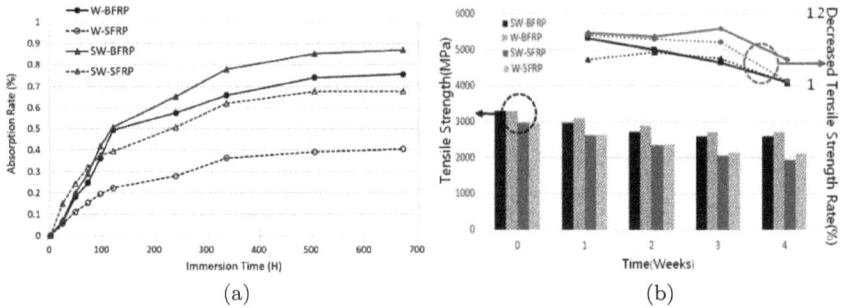

(a)   (b)

Fig. 4.30 Comparison of (a) water absorption rate and (b) tensile strength between BFRP and SFRP under fresh water/seawater environments (W: under fresh water, SW: under seawater).[18]

of interfacial erosion due to salinity in the seawater, despite the fact that seawater contained a larger amount of ions than water molecules compared in fresh water.

4.6.3   *Mechanical Properties and Failure Analysis Under Tensile Loading*

Figure 4.30(b) shows the change in tensile strength over time exposed to fresh water and seawater, and BFRP and SFRP commonly appeared with gradual strength deterioration in fresh water and seawater. BFRP has a relatively higher tensile strength than SFRP, and a low strength decrease rate was measured. Both BFRP and SFRP were more vulnerable to seawater than fresh water, and this is because salts and dissolved ions in seawater mainly cause polymer swelling and weaken the interfacial bonding force of the laminated composites. W-BFRP, however, maintained a low decreased strength rate despite the rapid inflow of large amounts of water based on strong interfacial bonding between epoxy and basalt fiber in neat BFRP.

4.6.4   *Morphological Structure Observation*

Figure 4.31 shows the fiber surface exposed at the composite interface through SEM, when the moisture absorption saturation is reached. A large amount of residues were commonly observed on the surface of

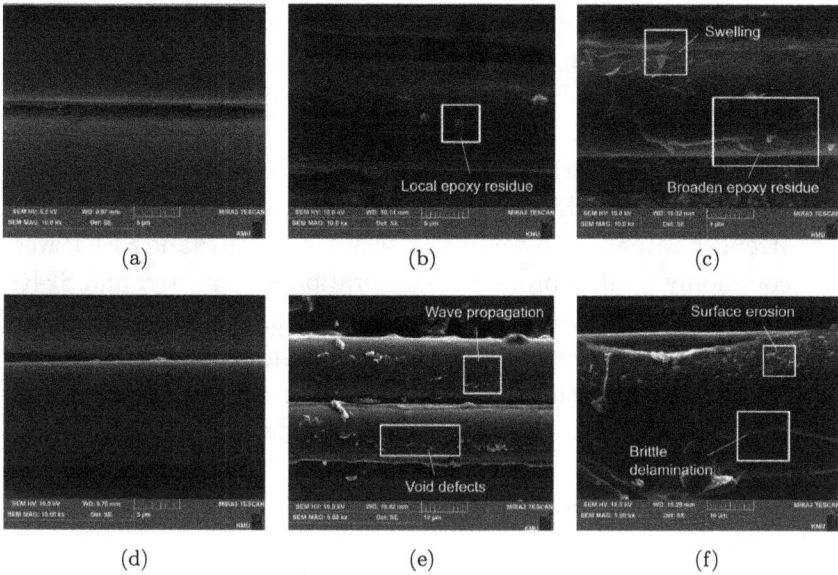

Fig. 4.31 Surface morphological changes through SEM observation; (a) Neat BFRP, (b) W-BFRP after 4 weeks, (c) SW-BFRP after 4 weeks, (d) Neat SFRP, (e) W-SFRP after 4 weeks and (f) SW-SFRP after 4 weeks.

SW-BFRP and SW-SFRP, and distributed along the interface with the torn epoxy. It is considered that not only moisture but also ions contained in seawater are directly related to the deterioration. In particular, swelling and fatal damages such as crack growth occurred in W-SFRP and SW-SFRP, which have relatively little moisture absorption, and delamination was noticeable at the fiber–epoxy interface in SW-SFRP. And it demonstrates lower tensile strength than BFRP.

4.6.5 *Conclusion*

(1) SFRP has a relatively superior moisture absorption resistance compared to BFRP, and exhibits a significantly lower moisture permeability, which increases moisture absorption resistance by up to 67% in conditions where interfacial erosion is not active, such as fresh water.

(2) The damage to the fiber surface was a major factor ultimately affecting moisture absorption resistance. This was involved in

promoting the acceleration of the initial moisture absorption due to the increase in the water permeation rate, but it did not affect the time taken to reach the final moisture saturation in the composites by moisture absorption.

(3) BFRP has a relatively high interfacial bonding strength compared to SFRP, which ultimately affects the low decreased tensile strength rate in fresh water and seawater. BFRP and SFRP were commonly used to promote deterioration in seawater, and SFRP was particularly vulnerable even to amounts of moisture.

(4) Customized slag fiber showed high utilization value as industrial fiber by deriving the moisture absorption resistance corresponding to basalt fiber in extreme water environments. However, optimization of the design process for fiberization of slag is required to improve durability.

4.7 Conclusions

This chapter aims to establish optimized process conditions for recycling steel slag, an industrial by-product, and to confirm the applicability of industrial fiber. Since the formation of industrially available ceramic fibers is very diverse, it is possible to diversify raw materials and products and to develop eco-friendly new fibers. By developing new fibers, it is possible to reduce the disposal cost of steel slag and develop high value-added materialization technology using it. In addition, unlike the recovery of metals or resources from existing single-creation waste resources, the technology that can present a new method of recycling through a combination of various process by-products was determined through the analysis of raw materials, research on the spinning processes, taking into account the hot viscosity characteristics of slag and the characteristics of fibers.

Based on the chemical composition ratio of basalt fibers suitable for roving fibers, raw materials mixed with fly ash, Fe–Ni slag and furnace slag at 40:30, 50:25:25 and 60:20:20 were manufactured through XRF analysis. After the primary melting of the raw materials, spinning of the fibers was rapidly cooled using water, and all of them made glassy materials without any crystalline parts. The analysis of XRD confirmed the amorphosis of raw materials. The material, which was identified as an atypical, was kept at the fibrous temperature for about 2 h, then the winding speed was adjusted and spinning was performed.

Through the viscosity model, the combination ratio of the resulting fly ash, Fe–Ni slag and electric furnace slag in the fibrous temperature with viscosity values of 2.8–3 was found to be the most stable, and stable spinning was conducted under the conditions of 1221°C and 16 m/sec.

The proportion of blended slag was 2.98, similar to that of glass fiber and basalt fiber. Water absorption rate, resistance to acid and abrasion properties all showed excellent characteristics. The tensile strength of the spinning slag fiber was similar to that of basalt and glass fibers. It has an average strength of about 3,000 MPa, and it is believed that the variation in fiber tensile strength was caused by defects such as blisters in the fiber or by variables of fine crack and stress in the cooling process.

It can be seen that the reduction in mass in NaOH and the decrease in tensile strength are smaller than when the HCl solution is used for submerging. SEM analysis also shows that the amount of residue on the surface of the fiber is higher, so slag fibers are thought to be superior in their chemical resistance in the NaOH environment to the HCl environment.

BFRP has a high interfacial bonding strength as compared to SFRP, which ultimately affects the low tensile strength reduction rate of fresh water and seawater. BFRP and SFRP were commonly used to promote seawater deterioration, and SFRP was particularly susceptible even to amounts of moisture. Customized slag fibers were found to be highly utilized as industrial fibers by deriving moisture absorption resistance corresponding to basalt fibers in water environments.

References

[1] B. Wei, H. Cao and S. Song, *Mater. Des.*, 31, 2010, 4244.
[2] C. W. Park, S. W. Yoon, S. J. Park and Y. H. Kim, *Mod. Phys. Lett. B* 33, 2019, 1940027.

[3] C. W. Park and Y. H. Kim, *Fiberization and Reuse of Slag for High Added Value and Its Application, in International Conference on Physics and Mechanics of New Materials and Their Applications*, Springer, 2017, pp. 589–603.

[4] F. Wang and J. Shao, *Polymers*, 6, 2014, 3005.

[5] H. J. Park, S. M. Park, J. W. Lee, G. C. Roh and J. K. Kim, *Compos. Res.*, 23, 2010, 43.

[6] J. S. Lee, T. Y. Lim, M. J. Lee, J. H. Hwang, J. H. Kim and S. K. Hyun, *J. Korean Cryst. Growth Cryst. Technol.*, 25, 2015, 263.

[7] K. Marko, O. Haile and L. Seppo, *Aalto University Publication Series Science+Technology*, 12, 2012, 1.

[8] L. C. Pardini and L. G. B. Manhani, *Mater. Res.*, 5, 2002, 411.

[9] R. Roy, B. K. Sarkar and N. R. Bose, *Bull. Mater. Sci.*, 24, 2001, 87.

[10] T. H. Lee, J. Y. Kim, K. Y. Kang, H. H. Cho and H. S. Kim, *T. Sci. Eng.*, 45, 2008, 262.

[11] V. D. Eisenhuttenleute, *Slag Atlas*, Verl. Stahleisen, 2012.

[12] V. Sinelnikov and D. Kalisz, *Arch. Foundry Eng.*, 15, 2015, 119.

[13] W. R. Broughton and M. J. Lodeiro, Techniques for Monitoring Water Absorption in Fibre-Reinforced Polymer Composites, Technical Report, United Kingdom, 2000.

[14] X. Jiu and Q. Wu, *Eur. Polymer J.*, 38, 2002, 1383.

[15] Y. Chen, S. Mori and W. P. Pan, *Thermochim. Acta.*, 275, 1996, 149.

[16] Y. Huh, H. J. Kim, H. W. Yang and K. J. Jeon, *J. Korean Soc. Precision Eng.*, 26, 2009, 78.

[17] Y. V. Lipatov, S. I. Gutnikov, M. S. Manylov, E. S. Zhukovskaya and B. I. Lazoryak, *Mater. Des.*, 73, 2015, 60.

[18] S. Y. Kim, S. J. Park, C. W. Park and Y. H. Kim, *Int. J. Mod. Phys. B*, 34, 2020, 2040130.

Chapter 5

Fatigue Strength of Fiber Reinforced Composites Made of Carbon Fibers, Glass Fibers and Other Fibers

Ri-Ichi Murakami

Chengdu University, School of Mechanical Engineering, Sichuan, China

anewmoon816@gmail.com

Compared to metal materials, because the continuous fiber reinforced composites of thermosetting resins (CFRTS) are higher in both specific strength and specific rigidity, they can be expected to be significantly lighter. While continuous fiber reinforced composites of thermoplastic resins (CFRTP) are not much different from metal materials in strength and have excellent specific rigidity, it can be seen that CFRTP is highly likely to contribute to the weight reduction of cars as automobile components if the excellent workability is utilized. This chapter describes the fatigue strength or fatigue fracture of CFRTP or related composite materials.

5.1 Introduction

Thermoplastic resin has advantages such as high reliability because it can be easily molded by heating and has high toughness and strength in terms of mechanical properties. The continuous fiber reinforced composites of thermoplastic resins (CFRTP) is a thermoplastic composite material that has high productivity because it can be molded in a short period of time due to the high molding temperature. The Japan Machinery Federation (JMF) and the R&D Institute of Metals and Composites for Future Industries (RIMCOF) investigate and report on the latest trends in materials technology, material development and component manufacturing equipment, as well as applicable technologies for aircraft parts, focusing on the aircraft field as an area where the characteristics of thermoplastic composite materials can be utilized.[1] Recently, high-speed and low-cost molding and processing techniques by which it is possible to produce automotive parts have been developed using the excellent features of CFRTP. As a result, carbon fiber composite materials are increasingly being used for automotive parts.[2]

In the project of the NEDO, the University of Tokyo, Toray Industries, Mitsubishi Rayon, Toyobo Co., Takagi Seiko and so on have studied the effects of molding conditions, fiber content and cavity ratio on mechanical properties, and have succeeded in developing CFRTP, which enable high-speed molding and highly versatile bonding.[3] This technology is expanding to the application to mass production vehicles.

In order to consider CFRTP as a structural member of an automobile, Fig. 5.1 shows the comparison between the specific strength and specific stiffness of CFRTP and other structural members.[4] For metallic materials, as the type of material is changed, the specific

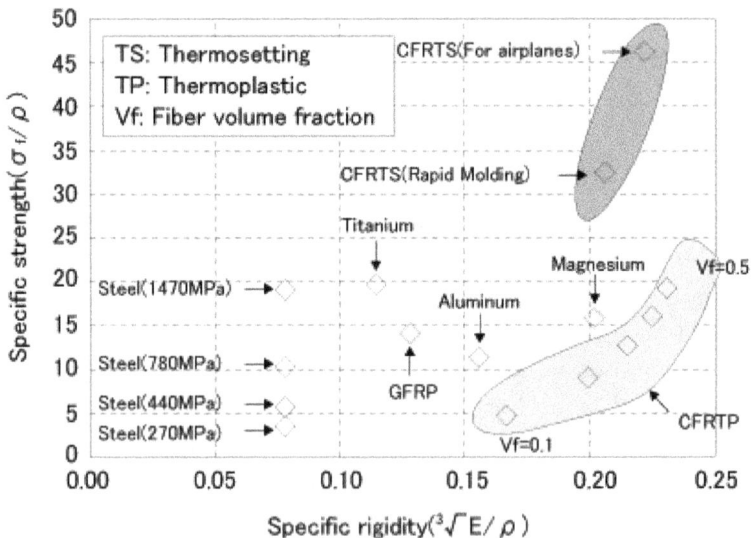

Fig. 5.1 Comparison of specific strength and specific stiffness of metal and composite materials.

strength changes, but the specific stiffness tends to be constant. However, for composite materials, specific strength and specific stiffness depend greatly on fiber morphology, fiber content and molding method. Because thermosetting composites (CFRTS) are greater for both specific strength and specific rigidity than metallic materials, they can be expected to be significantly lighter than metal materials. The strength of thermoplastic composites (CFRTP) is almost the same as that of metal materials, but they have excellent specific rigidity and can contribute to reducing the weight of vehicles as automobile parts by taking advantage of their easy workability.

In this chapter, for CFRTP or other reinforcing fiber composites, the effects of fiber content, fiber orientation, humidity and temperature, and the frequency rate of fatigue strength and fatigue crack growth rate described are based on the latest research results.

5.2 Comparison of Fatigue Properties of CFRTP and Aluminum Alloy

Although it is generally known that CFRTP has excellent fatigue strength characteristics, it differs considerably from metallic materials. Figure 5.2 compares the tensile–tension fatigue properties of an CFRTP plate with a circular hole to that of an aluminum alloy.[5] CFRTP has a higher fatigue strength than aluminum alloys, especially in the long life range. This is because the carbon fiber, which is the reinforcing fiber of CFRTP, scarcely has fatigue damage, it is characterized for the S–N curve by almost horizontal nature. The fatigue life in CFRTP is not determined by fatigue crack initiation, but most of the fatigue life is occupied by fatigue crack growth; carbon fibers oriented in CFRTP are remarkably effective in deterring fatigue crack growth.

5.3 Effects of Frequency Rate and Temperature on Fatigue Strength of Composite Materials

Dumbbell-type (DB) specimens are used to test the fatigue strength of carbon fiber reinforced composites. For carbon fiber reinforced

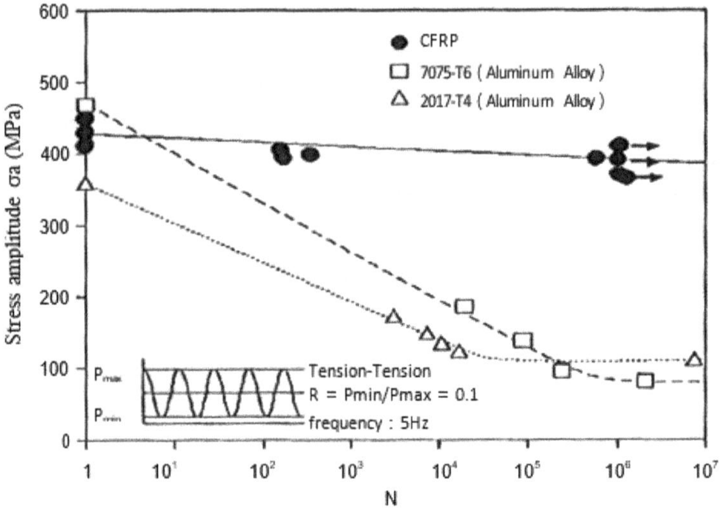

Fig. 5.2   Comparison of tensile fatigue properties of aluminum alloy and CFRTP with hole notch.

polyphenylene sulfide composite (CF-PPS) using DB Test Specimens, the tension–tension S–N curves are shown in Fig. 5.3.[6] The S–N curve for a rectangular specimen is also shown in Fig. 5.3. The DB specimens have a stress concentration factor of 1.1–1.2, but in the maximum stress above 675 MPa, there is no significant difference in fatigue life when the frequency is increased from 2 Hz to 5 Hz. However, below the stress of 675 MPa, the fatigue life decreases with increasing frequency. On the other hand, the results for the rectangular specimens are different from those for the DB specimens. By the specimen with or without stress concentration, the effect of frequency on fatigue life results in significant differences.

Figure 5.4 shows the ultra-long life S–N curve of CF-PPS in the three-point bending fatigue test at a frequency of 20 kHz.[7] The failure defined by macroscopic delamination occurred at 6.0×105 cycles under a shear stress amplitude of $\tau a.13$ of 8.0 MPa. When the shear stress amplitude at 109 cycles was less: $\tau a.13 = 5.5$ MPa or over 4.5 MPa, the fatigue failure was observed. But from 4 MPa to 4.5 MPa, no fatigue damage was observed. While, at $\tau a.13 = 4$ MPa,

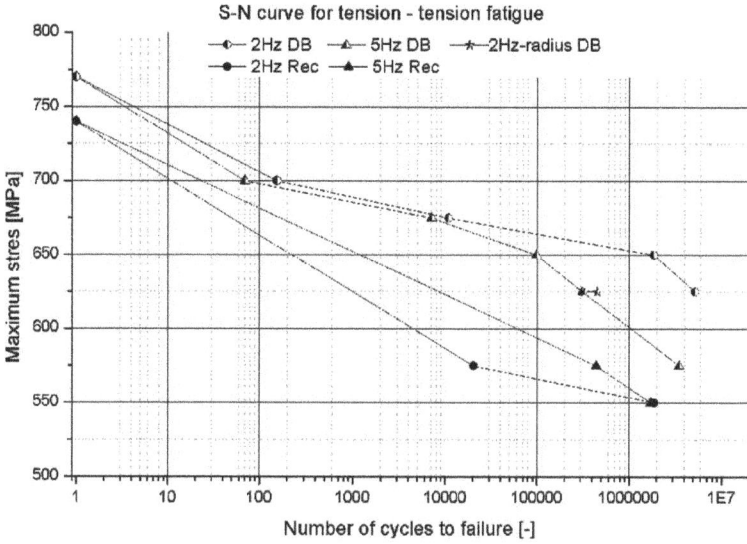

Fig. 5.3 S–N curve for tensile fatigue at frequency of 2 Hz and 5 Hz.

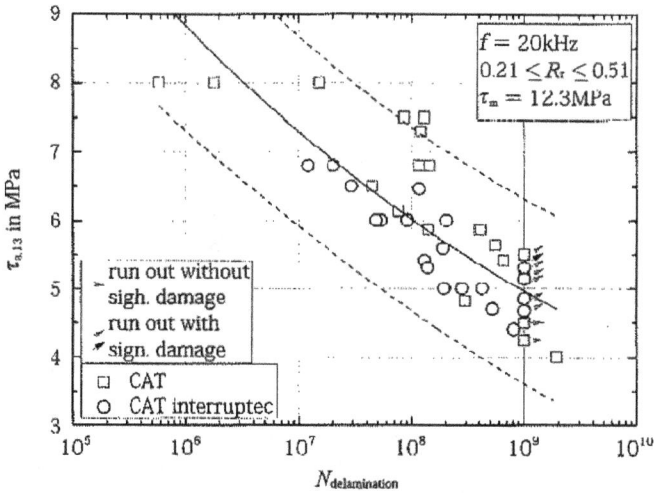

Fig. 5.4 S–N curve of CF-PPS at frequency of 20 kHz.

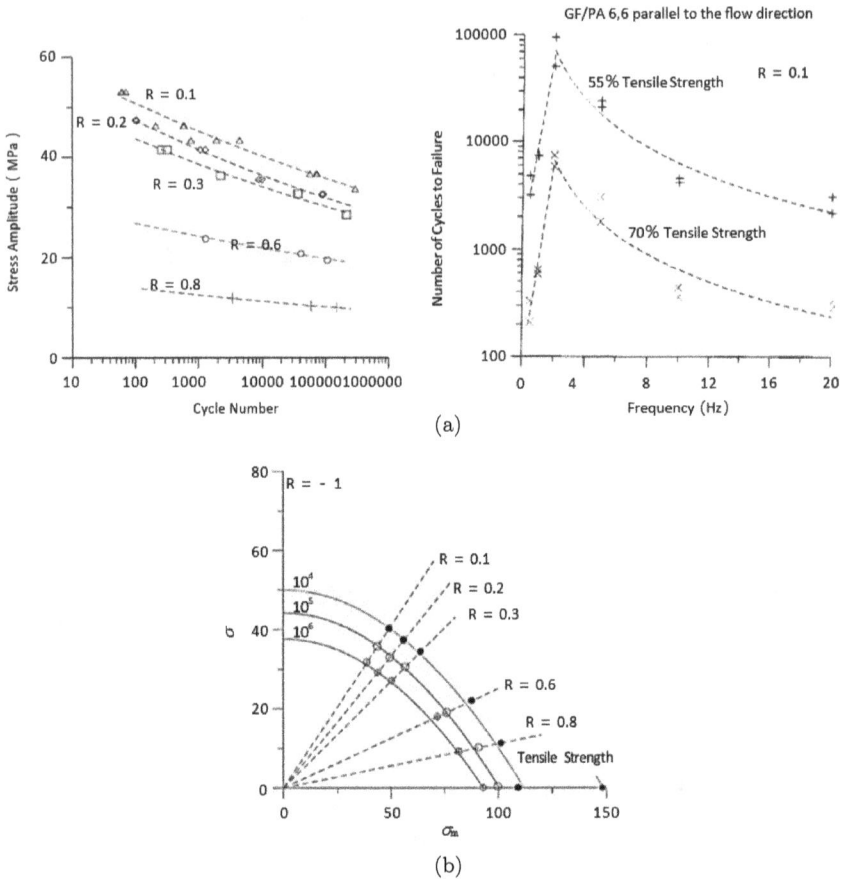

Fig. 5.5 Effect of frequency on the fatigue life of short glass fiber reinforced polyamide-6.6.

fatigue failure at 1.91×109 cycles had occurred, the same as the other specimens.

Figure 5.5 shows the effect of frequency on the fatigue life of short glass fiber reinforced polyamide-6.6 composites.[8] Fatigue life increases with frequency up to 2 Hz under each stress level. After that, when the frequency increases, the fatigue life decreases. This results from the fact that the average strain increases with the number of cycles under a frequency of 2 Hz, whereas the specimen temperature increases above 2 Hz due to hysteresis heating. Such

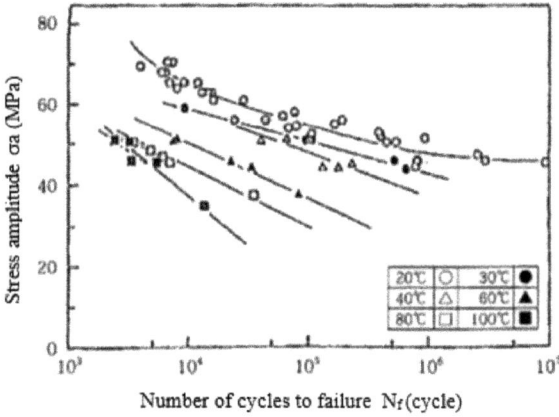

Number of cycles to failure N_f(cycle)

Fig. 5.6 Effect of test temperature on fatigue strength (PAN-based short fiber reinforced Polyacetal composite, carbon fiber content of 30%).

hysteresis heating is caused by plastic or viscoelastic deformation in the matrix phase. Bellenger *et al.*[9] showed that the specimen temperature increase at 2 Hz was 10°C, but at 10 Hz, the temperature increase was 100°C. Figure 5.6 shows the S–N curves for Polyacetal CFRTP with 30% carbon fiber content when the test temperature was changed to 20°C, 30°C, 40°C, 60°C, 80°C and 100°C.[10] The fatigue strength decreased rapidly with increasing test temperature T. The fatigue strength increases in proportion to the inverse of the test temperature, regardless of the number of cycles. In the range of 20–100°C, because carbon fibers are scarcely affected by temperature, the decrease in fatigue strength results from an increase in viscous flow of Polyacetal with an increase in the temperature. Thus, the increase of test temperature results in the same effects as the frequency, and results from the decrease in the fatigue life.

Figure 5.7 shows the relationship between the crack growth rate and the frequency for CF-PEEK at 473 K.[11] As shown in Fig. 5.7, the crack growth rate divides into the dependence on the stress cycle or time at $1/\nu = 20$ s. That is, at high temperatures, it is clear for CF-PEEK that the crack growth rate divides into either fatigue damage or creep damage depending on the frequency. In the case of predominantly fatigue damage, the crack growth rate increased with

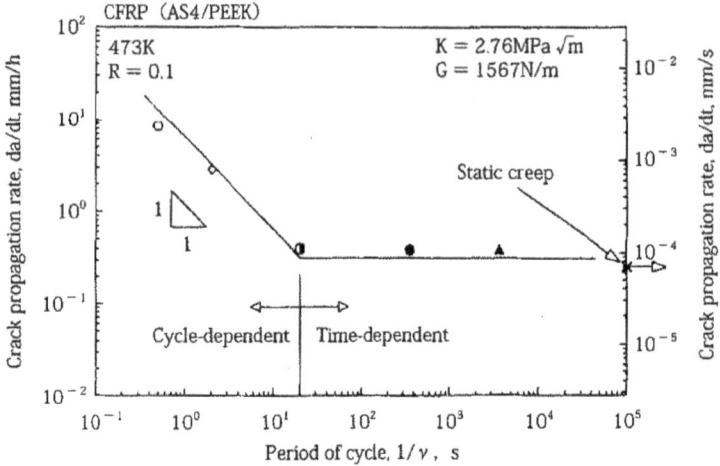

Fig. 5.7 Relationship between period of cycle $(1/\nu)$ on fatigue crack growth rate at $K = 2.76\,\mathrm{MPam}^{1/2}$.

increasing temperature, as shown in Fig. 5.8. Similar results were obtained by Tanaka et al. on short carbon fiber reinforced PPS.[12]

5.4 The Effect of Notch on the Fatigue Strength of Composite Materials

Figure 5.9 shows the S–N curve of CFRTP when the notch radius is changed to 0.4, 1.2, 2.0 and 40 mm.[10] As can be seen from Fig. 5.9, the fatigue strength of CFRTP is scarcely affected by the notch radius in the low fatigue life range, but in the high fatigue life range, the fatigue strength decreases as the notch radius decreases. Figure 5.10 shows the effect of notch radii on the fatigue strength at room temperature. As shown in Fig. 5.10, the results of glass short fiber reinforced polyamide-6.6 composite materials[13, 14] were similar to that of CFRTP. It is the tendency at the high temperature of 130°C that the difference in fatigue strength between smooth and notched specimens decreases.

In general, tensile strength depends on the notch geometry and is governed by the maximum stress at the notch tip. On the other hand, it is well known that the fatigue strength of metals is sensitive to their

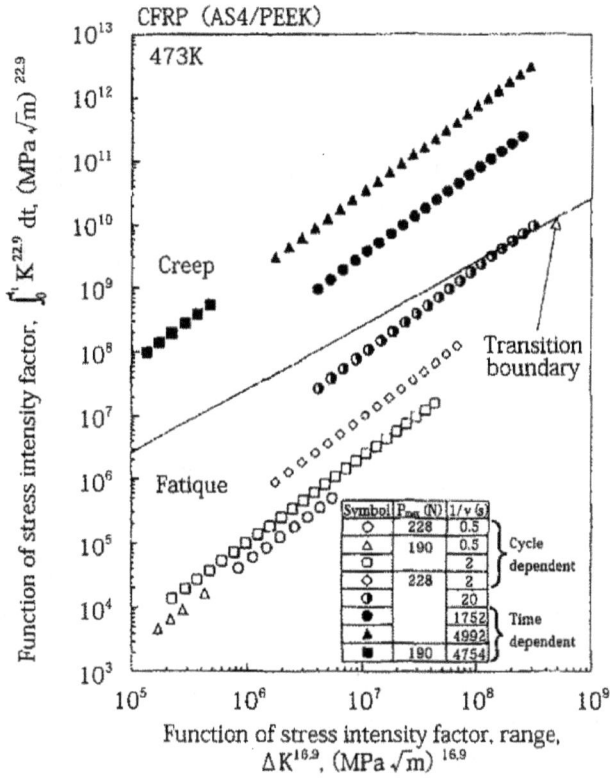

Fig. 5.8 Effects of period of cycle $(1/\nu)$ and test temperature on fatigue crack growth rate.

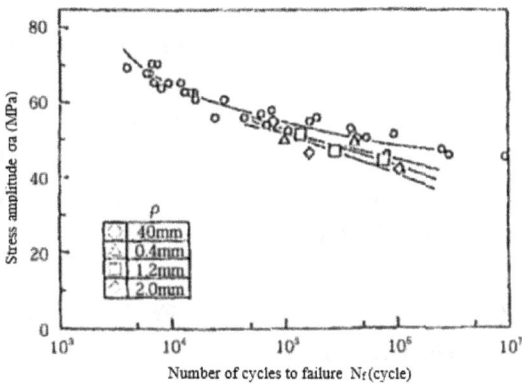

Fig. 5.9 Effect of notch radius on fatigue strength, o:40mm of ρ, Δ:0.4mm of ρ, $\square$:1.2mm of ρ and $\diamond$:2.0mm of ρ.

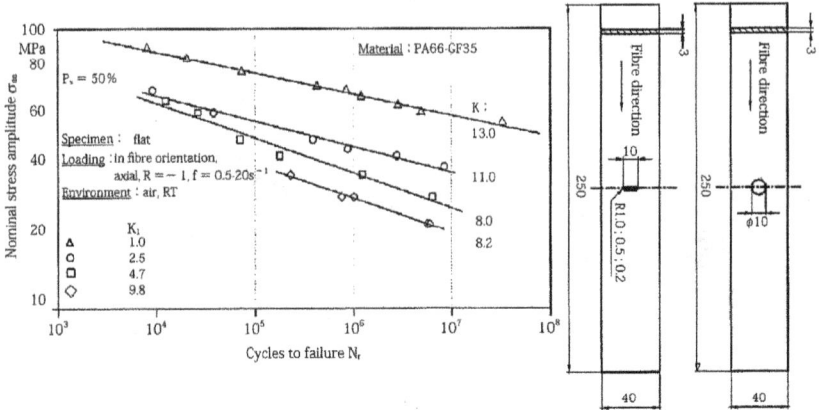

Fig. 5.10   Effect of stress concentration factor on fatigue strength of short glass fiber reinforced polyamide-6.6.

microstructure.[15] However, in composites with randomly oriented short fibers, because the carbon fibers constrain the deformation of the matrix phase, the stress concentration occurs at the fiber ends which are discontinuities in the fibers, and at the places where the fibers cross each other. Thus, the composite materials result in microscopic stress unevenness. When such a composite material is notched, as the temperature of the test increases, the ductility of the matrix increases with an increase in the test temperature. Because in composite materials notch sensitivity is blunted, the fatigue damage of composite materials with increasing temperature changes from a system in which nominal stresses mainly cause damage to a system in which local stresses are the main cause of damage. Therefore, the fatigue strength of the composite material is greatly influenced by the microscopic sources of stress concentration inside the composite materials, and the effect of stress concentration due to mechanically introduced notches is reduced.[16]

5.5   Effect of Fiber Orientation on Fatigue Strength of Composite Materials

The fatigue strength of the composite material is affected by the ply orientation of the reinforcing fibers. Figure 5.11 shows the S–N curve of long carbon fiber reinforced bioplastic composites in [0°] and [45°] orientation, made of Ecoflex which has cornstarch as the main ingredient and is blended with PLA.[17] The fatigue strength of composites with carbon fibers in [0°] orientation is greater than that of composites in [45°] orientation. As similar to the tensile strength, the tensile stress is mainly borne by the carbon fibers of [0°] orientation, whereas for composites oriented at [45°] orientation, the matrix phase near the interface is also damaged by deformation of the plain-woven carbon fiber. The results shown in Fig. 5.11 suggest that because the fatigue damage of composites oriented at [45°] enlarged, the fatigue strength was reduced. However, the difference between the two decreases in the long fatigue life range at low stress level. As the stress amplitude decreases, in [45°] orientation composites, in addition to the carbon fibers, many other factors like the interfaces and intersections of the carbon fibers can bear the stress. The effect of fiber orientation on fatigue strength is likely to be lessened, as [45°] orientation composites have equivalent effect of carbon fibers being mainly stressed in composites with [0°] orientation.

Horst and Spoormaker conducted tensile–tension fatigue tests on short glass fiber polyamide-6 and showed that the fatigue strength

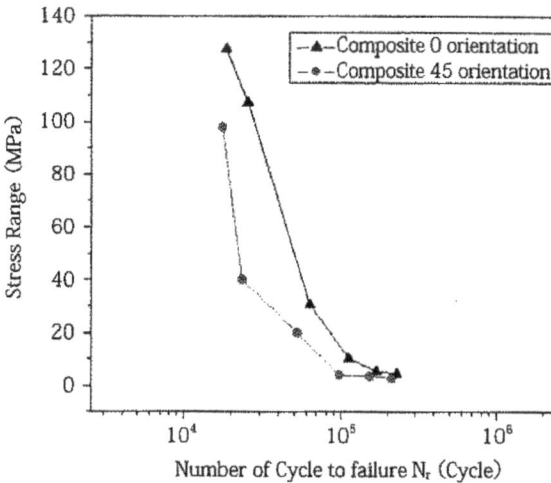

Fig. 5.11 Effect of fiber orientation on fatigue strength of long carbon fiber reinforced bioplastic composites (cornstarch with PLA).

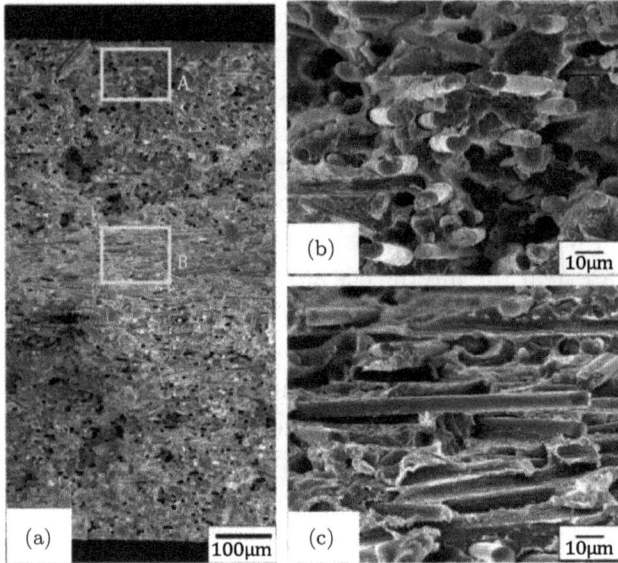

Fig. 5.12 SEM fracture appearances of fatigue fracture surface of MD-IMP at R = 0.1; (a) fracture surface appearance, (b) microscopic fracture appearance of subsurface and (c) microscopic fracture appearance of B part at core layer.

depends greatly on the position of the specimens collected.[18] In other words, the fatigue strength of specimen taken at the center of the injection molded specimen was 50% higher than that of specimen taken from the edge. This is due to the fact that the orientation of the fibers differs depending on the position of the specimen during injection molding. This can be seen in the electron micrograph of the fatigue fracture surface shown in Fig. 5.12.[19] The shell layer near the surface of the injection molded specimen is in the orientation of the fibers along the resin flow during injection molding. In contrast, most of the fibers in the core layer are oriented perpendicular to the flow of the resin.[20]

Figure 5.13 shows the effect of fiber orientation on the fatigue crack growth rate for stress ratios of 0.1 and 0.5.[19] Specimens cut along the direction of injection are at [0°], cut at 45° are at [45°] and vertically cutout are at [90°]. For all stress ratios, as the direction of cutout changes from 0° to 90°, the fatigue crack growth rate

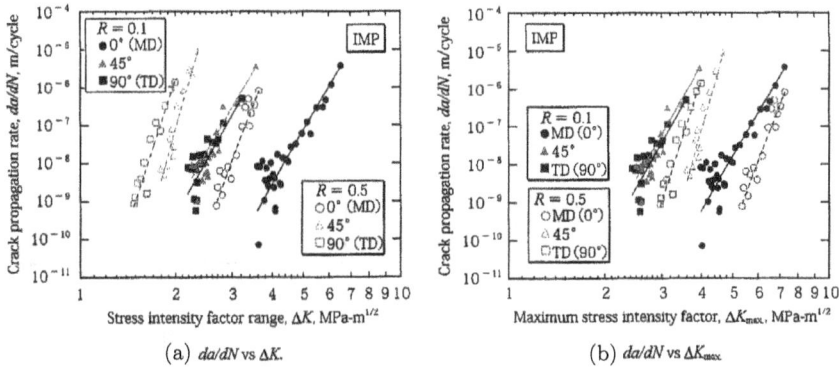

Fig. 5.13 Effect of specimen orientation on fatigue crack growth rate of IMP at R = 0.1 and 0.5; MD: specimen orientation of 0° means the parallel to molding flow direction and TD: specimen orientation of 90° means the perpendicular to molding flow direction.

increases. This suggests that the resistance of the fibers to crack growth differs by the orientation of the fibers with respect to the direction of crack growth. Similar results for fatigue strength can be confirmed from Fig. 5.14 where the effect of the cutout angle of the specimen from the injection molded specimen was investigated. As the cutout angle of the specimen increases, the orientation of the fibers reduces fatigue strength because the fiber changes from 0° to 90° against the stress axis.[21]

5.6   The Effect of Fiber Content on the Fatigue Strength of Composite Materials

Figure 5.15 shows the S–N curve of CFRTP with 0, 15 and 30% PAN-based short fibers in Polyacetal, which is a type of thermoplastic resin.[10] Since Polyacetal did not fail except when the stress amplitude was quite low, fatigue failure was defined as the point at which the displacement of the specimen exceeded 8 mm, which is the permissible range of the fatigue testing machine. In Fig. 5.15, these results are indicated by the symbol of □. This is the same for the CFRTP with 15% CF content. However, when the content was increased to 30%, the CFRTP failed under all cyclic stresses. At a

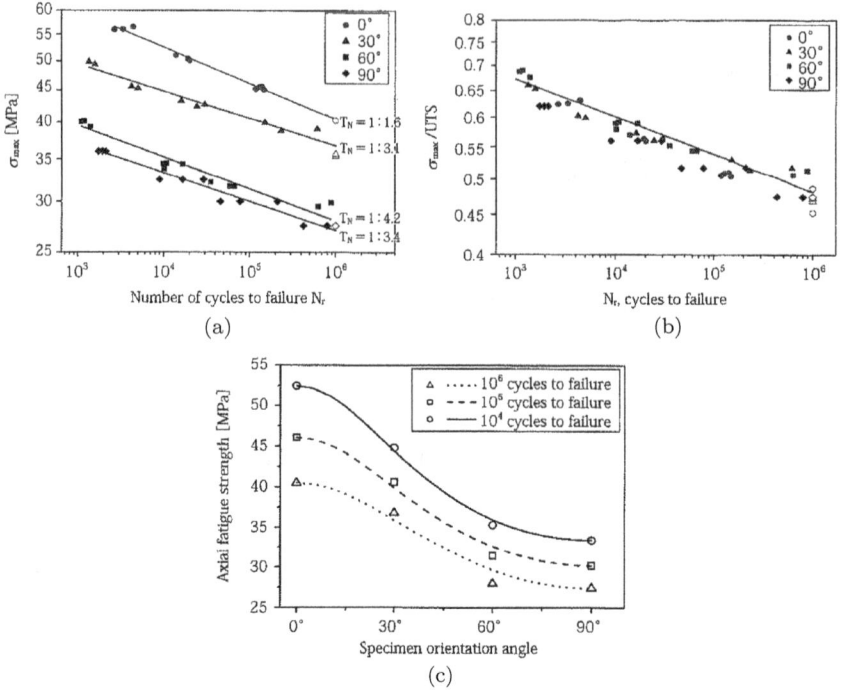

(a)

(b)

(c)

Fig. 5.14 Fatigue strength of a 3.2 mm thick specimen of short glass fiber reinforced polyamide-6 at R = 0.1 one = sided test relationship.

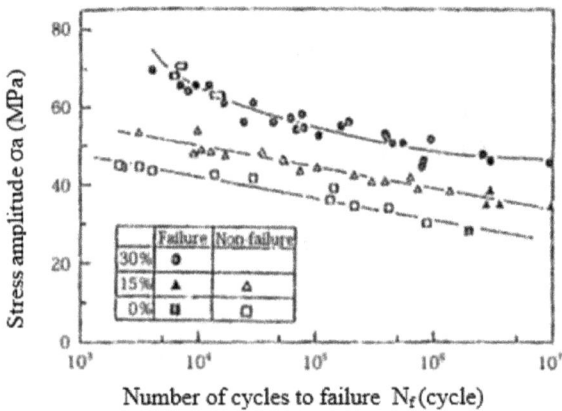

Fig. 5.15 Relationship between fatigue strength and short carbon fiber content (short PAN-based carbon fiber reinforced Polyacetal composites, carbon fiber content of 0.15% and 30%).

specific number of cycles, as the carbon fiber content increased, the fatigue strength of Polyacetal CFRTP increased almost linearly. This implies that the same compounding law as tensile strength holds.

5.7 Effects of Humidity and Chemical Aging on Fatigue Strength of Composite Materials

For carbon fiber reinforced composites oriented by carbon fiber within Epoxy and PEEK, the effect of humidity on fatigue strength is shown in Fig. 5.16 when the fatigue test was conducted under the conditions of R = 0.1 and f = 10 Hz.[22] For each specimen, fatigue tests were stopped at 2×106 cycles. In a dry environment, as shown in Fig. 5.16(a), fatigue strength depends on the type of matrix phase. For PEEK, the fatigue strength is the lowest in the case of this study. Under saturated humidity, the differences in fatigue strength due to the matrix phase are reduced, as shown in Fig. 5.16(b). In other words, the fatigue strength of epoxy is reduced by 7–15%, while PEEK is scarcely affected by humidity. The effect of humidity on tensile strength is not significant for composites in [0°] orientation. When in [90°] orientation, the tensile strength of epoxy-based composites is reduced with increasing humidity, while that of PEEK is almost unaffected. When the matrix phase is epoxy, the glass transition temperature decreases significantly with

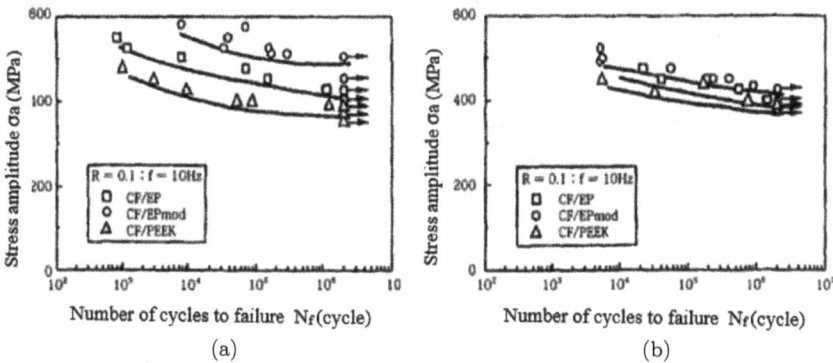

Fig. 5.16 Effect of humidity on the S-N curve of plain-woven CF/PEEK composites; (a) dry environment and (b) saturated humidity environment.

increasing humidity, but that is scarcely changed in PEEK. Thus, the
characteristics of epoxy matrix can be significantly changed as the
matrix and interface absorb water. Therefore, the fatigue strength
of the epoxy-based composites was strongly affected by humidity,
whereas it was considered that for the PEEK-based composites,
the fatigue strength was less affected by humidity because the glass
fiber temperature of PEEK was hardly affected by humidity. Short
glass fiber reinforced epoxy composites were aged at 180°C for 0 h,
250 h and 2000 h. When the fatigue test was conducted at room
temperature (23°C), with R = 0.1 and f = 10 Hz,[23] the S–N curves
of the virgin material and aging material of 250 h and 180°C were
almost identical. However, because the slope of the S–N curve of the
2000 h and 180°C aging material decreases, it is the tendency that
the fatigue life increases at low stress amplitudes.

5.8 Electron Microscopic Observation of Fatigue
Fracture Surface of Composite Materials

Figures 5.17 and 5.18 show electron micrographs of fatigue fracture
surfaces of long carbon fiber reinforced bioplastic composites in which
long carbon fibers were in [0°] and [45°] orientation to the matrix
phase blended with Ecoflex and PLA.[17] As shown in Fig. 5.17,
it is clear that the fracture surface of composites with carbon
fibers oriented along [0°] is clearly different from those with [45°]
orientation. Because carbon fibers with [0°] orientation are parallel
to the stress axis, the rupture of carbon fiber and the delamination of
the interface are seen. For composites in [45°] orientation, the bundles
of plain-woven carbon fibers are deformed and it appears that the
bundles have been pulled out and broken. These fractured surface
appearances indicate that the fatigue damage varies with the fiber
orientation.

Figure 5.19 shows the microscopic appearance of fatigue fracture
surfaces in shell layers of the MD ($\theta = 0°$) and TD ($\theta = 45°$) for
short carbon fiber reinforced PPS when fatigue test was conducted
at room temperature and 403 K.[24] As shown in Fig. 5.17, the fatigue

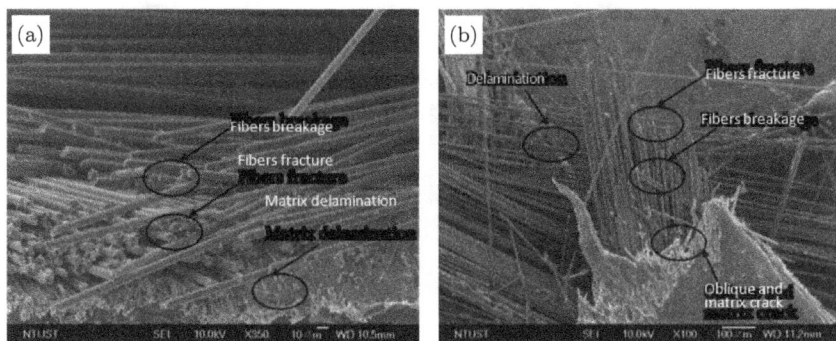

Fig. 5.17 Long carbon fiber reinforced bioplastic composites (containing PLA in cornstarch). Microscopic aspect of fatigue fracture surface of fiber orientation [0°]. (a) cross section area and (b) surface area.

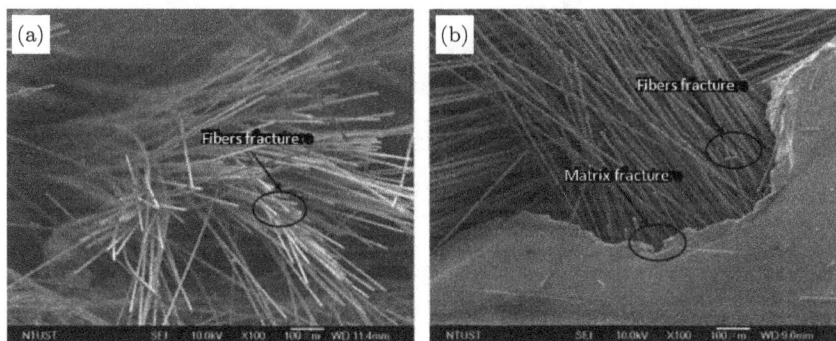

Fig. 5.18 A long carbon fiber reinforced bioplastic composite (containing PLA in cornstarch). Microscopic aspect of fatigue fracture surface of fiber orientation [45°]. (a) cross section area and (b) surface area.

fracture surface in the MD shows that the fibers are pulled out of the matrix phase, which is flat but heavily deformed. In the TD, the fibers lying parallel to the fracture surface can be seen after the delamination of the interface. Therefore, the observation of the fatigue fracture surface by electron microscopy can deeply elucidate the effects of fiber type and orientation, humidity and temperature on fatigue failure of the composite materials.

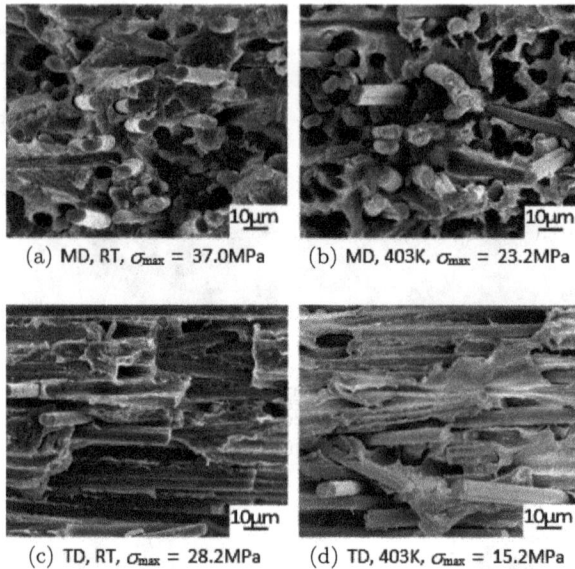

(a) MD, RT, σ_{max} = 37.0MPa   (b) MD, 403K, σ_{max} = 23.2MPa

(c) TD, RT, σ_{max} = 28.2MPa   (d) TD, 403K, σ_{max} = 15.2MPa

Fig. 5.19 Microscopic aspects of fatigue fracture surfaces of MD and TD shell layers subjected to fatigue tests at room temperature and 403 K.

5.9 Conclusions

(1) CFRTP has a higher fatigue strength than aluminum alloys, especially in the long life range. This is because the carbon fiber, which is the reinforcing fiber of CFRTP, scarcely has fatigue damage, it is characterized by an S–N curve that is almost horizontal in nature.

(2) The fatigue life of carbon fiber reinforced polyphenylene sulfide composites (CF-PPS) was not significantly affected by increasing the frequency from 2 Hz to 5 Hz at high stresses. However, at lower stresses, the fatigue life decreased with increasing frequency. For the short glass fiber reinforced polyamide-6.6 composites at all stress levels, the fatigue life increased with an increase in the frequency up to 2 Hz, but over 2 Hz the fatigue life decreased with the increase in the frequency. For CFRTP with 30% carbon fiber content, the fatigue strength increased rapidly with increasing test temperature. The fatigue crack growth rate

at 473 K for CF-PEEK was divided into two categories by $1/\nu = 20$s: in the less than 20 s of $1/\nu$, the stress showed cycle dependence, while over 20 s, it showed time dependence.

(3) The fatigue strength of CFRTP is largely unaffected by the notch radius in the low cycle range, but in the high cycle range, fatigue strength decreased with the decrease in the notch radius. When fatigue tests were conducted at 130°C, the difference in fatigue strength between smooth and notched materials was less than that at room temperature.

(4) For long carbon fiber reinforced bioplastic composites, the fatigue strength of composites with carbon fibers oriented in [0°] direction was greater than that of composites oriented in [45°] direction. In glass fiber reinforced polyamide-6 composites, the fatigue strength varied greatly, depending on the location of the specimens. Fatigue crack growth rates for stress ratios of 0.1 and 0.5 were determined by the orientation of the fibers with respect to the crack growth direction. That is, the results differed greatly depending on whether the crack growth was parallel to the fiber direction or not.

(5) For CFRTPs containing 0%, 15% and 30% PAN-based short fibers in Polyacetal, the fatigue strength increased with increasing carbon fiber content, and a similar compound law as the tensile strength was observed.

(6) Effects of humidity and chemical aging on the fatigue strength of composite materials: The fatigue strength of carbon fiber reinforced composites with epoxy and PEEK showed a general S–N curve with a decrease in stress amplitude and an increase in fatigue life. In dry environments, the fatigue strength depended on the type of matrix phase, and was the lowest for PEEK. In saturated humidity, the differences in fatigue strength by matrix phase decreased. For short glass fiber reinforced epoxy composites, when the composites were aged at 180°C for 0 h, 250 h and 2000 h, the fatigue strength of the no-aging and 250 h at 180°C aging materials were almost identical, while for the 2000 h at 180°C aging materials, the fatigue life in low stress amplitude tended to increase. Fatigue fractures of long carbon

fiber reinforced bioplastic composites blended with Ecoflex and PLA differed markedly between [0°] and [45°] orientations of the carbon fibers. Fatigue fracture surfaces of glass fiber polyamide-6 composites showed interfacial delamination, and in the unstable fracture surfaces, a hackle pattern, which indicates a breakdown of the interface and the matrix phase covered by the fibers, was observed.

References

[1] The Japan Machinery Federation (JMF) and R&D Institute of Metals and Composites for Future Industries (RIMCOF), 2008 Investigation Report on the Application of Thermoplastic Composites to Aircraft Field, March 2009.

[2] Matsunaga *et al.*, Development of Low-Cost CFRTP Processing Technology (Report 1), CFRTP Analysis Technology and the Effect of Molding Conditions on Mechanical Properties, Research Report of Seibu Industrial Technology Center, Hiroshima Prefectural Institute of Technology, No. 54, 2011, pp. 1–4.

[3] NEDO News Release, NEDO Technical Innovation Award to Carbon Fiber Composite Materials Development Project, http://www.nedo.go.jp/news/press/AA5_100130.html.

[4] A. Takahashi, Special Issue on Automotive Materials: Evolution and Challenges of Materials — Various Possibilities and Challenges of Automotive Materials from the Viewpoint of Recycling, Japan Automobile Manufacturers Association, Inc., http://www.jama.or.jp/lib/jamagazine/200603/05.html.

[5] Atsuhiko Kitano, *Chem. Edu.*, 59(4), 2011, 226–229.

[6] I. De Baere, W. Yan Paepegem, C. Hochard and J. Degriek, *Polym. Test.*, 30, 2011, 663–672.

[7] D. Backe and F. Balle, *Compos. Sci. Technol.*, 126, 2016, 115–121.

[8] Y. Zhou and P. K. Mallick, *Polym. Compos.*, 27, 2006, 230–237.

[9] V. Bellenger, A. Tcharkhtchi and P. Castaing, *Int. J. Fatigue*, 28, 2006, 1348–1352.

[10] R.-I. Murakami, M. Masuda and Toshio Nonomura, *J. Jpn. Soc. Mech. Eng.*, 58(545), 1992, 9–14.

[11] Y. Uematsu, T. Kitamura and R. Ohtani, *Compos. Sci. Technol.*, 53, 1995, 333–341.

[12] Keisuke Tanaka, Kazuya Oharada, Daiki Yamada and Kenichi Shimizu, Effect of test Temperature on Fatigue Crack Propagation in Injection Molded Plate of Short-fiber Reinforced Plastics, *Procedia Structural Integrity*, 2, 2016, 058–065.

[13] C. M. Socino, E. Moosbrugger, *Int. J. Fatigue*, 30, 2008, 1279–1288.

[14] S. Mortazavian and A. Fatemi, *Adv. Mater. Res.*, 891–892, 2014, 1403–1409.

[15] R.-I. Murakami, Y. Kim and K. Kusukawa, *Fundamental Strength and Fracture of Materials*, Fikuro Publishing, 2009, p. 74.

[16] A. K. Haldar and S. Senthilvelan, *Key. Eng. Mater.*, 471, 2011, 173–178.

[17] A. Fajrin, W. Solafide and R. Murakami, The effect of Fiber Ply Orientation, Laminate Layer and PLA content on Mechanical Properties of Carbon Fiber Reinforced Bio Plastic Composites, 2nd International Conference on Nanomaterials and Advanced Composites, August, 2019, Taipei, Taiwan.

[18] J. Horst and J. L. Spoormaker, *Polym. Eng. Sci.*, 36, 1996, 2718–2726.

[19] K. Tanaka, T. Kitano and N. Egami, *Eng. Fract. Mech.*, 123, 2014, 44–58.

[20] M. F. Arif, N. Saintier, F. Meraghni, J. Fitoussi, Y. Chemisky and G. M. Robert, *Compos. Part B, Eng.*, 61, 2014, 55–65.

[21] A. Bernasconi, P. Davoli, A. Basile and A. A. Fillipi, *Int. J. Fatigue*, 29, 2007, 199–208.

[22] R. Selzer and K. Friedrich, *Compos., Part A*, 28A, 1997, 595–604.

[23] D. Flore, K. Wegener, D. Seel, C. C. Oetting and T. Bublat, *Compos.: Part A*, 90, 2016, 359–370.

[24] Keisuke Tanaka, Kazuya Oharada, Daiki Yamada and Kenichi Shimizu, *Procedia Structural Integrity*, 2, 2016, 058–065.

Chapter 6

Corrosion and Tribological Properties of Basalt Fiber Reinforced Composite Materials

Jin-Cheol Ha[*,‡], Soo-Jeong Park[†,§] and Yun-Hae Kim[†,¶]

*College of Engineering, Dali University,
Yunnan, China
†Department of Ocean Advanced Materials
Convergence Engineering,
Korea Maritime and Ocean University,
Busan, Republic of Korea
‡chnhjc@naver.com
§blue9069@naver.com
¶yunheak@kmou.ac.kr

The chemical reaction, mass change, tensile strength and fiber corrosion mechanism in the alkaline solution of the basalt fiber itself were examined. Furthermore, the specimen was prepared by the Vacuum assisted resin transfer molding (VaRTM) method and exposed to an aqueous sulfuric acid solution to investigate the friction and wear characteristics. As a result, the basalt fibers were immersed in various alkaline solutions, and weight retention and tensile strength were observed to find out the chemical stability in alkaline solutions. It has been confirmed that weight retention is affected by the type of alkaline solution. However, regardless of the type of alkaline solution used, the tensile strength decreased significantly during the initial stage of immersion. Also, the weight retention of basalt fiber in 0.4% NaOH solution was much higher than that of a 10% NaOH solution. However, the decrease in tensile strength remained independent of concentration and still fell significantly. In particular, the corrosion behavior of basalt fiber according to the temperature in NaOH solution showed a different tendency of change in tensile strength as the corrosion time increased. At 70°C high temperature, the corrosion rate of the basalt fiber accelerated, increasing the mass loss rate of the basalt fiber as the temperature increased due to the leaching of silicon, aluminum and potassium ions. Looking at the progress of the friction coefficient of the specimen before corrosion, if the friction coefficient becomes constant after about 77.19 m, in the case of the specimen after corrosion, the longest distance is about 16.63 m, and the shortest distance is 3.02 m, at which the friction coefficient becomes constant.

6.1 Corrosion of Basalt Fiber Reinforced Composite Materials

Chemical corrosion of composites is inevitable for some applications. For example, some containers, ships, tubes, off-shore platforms and equipment in marine applications can become corrosive after prolonged use in an alkaline environment. One of the obstacles to the widespread use of composites is the lack of long-term durability and performance data when servicing in critical environments. Consequently, it is necessary to understand how the material behaves during long-term application.[1,2]

Most of the previous research on corrosion of reinforcing fibers in a chemical environment has focused on processed materials in which the fibers are covered with a matrix. The fiber's resistance to corrosion is as follows. It mainly depends on the corrosion resistance of the resin, and the corrosion crack propagation is related to the resin toughness. This is the reason why the underlying mechanisms of fiber damage and deterioration were not recognized. Damage to the fiber surface can be detected with prolonged exposure in critical environments.[3-5]

It is also seen that stressed fiberglass is susceptible to corrosive environments.[6,7] There have been many attempts to investigate the response of glass fiber reinforced products to critical environments;[8-13] however, many people did not care about the properties of the fibers themselves.[14]

Due to the long-term chemical resistance of mineral fibers, they cannot be explicitly measured or determined. So, in general, we are doing comparative experiments after accelerated aging under fixed experimental conditions. The rate of corrosion is determined by various parameters such as temperature, aging time, fiber composition, the composition of the aging solution and fiber size.[15] Wei *et al.* investigated the direct effects of NaOH and HCl on the performance filaments of basalt and glass fibers, excluding the influence of the matrix and revealing a clear and accurate response of the fibers to chemical corrosion.[16] Lee *et al.* studied the chemical stability of basalt fiber in alkaline solution that was investigated utilizing weight retention and tensile strength retention.[17] In recent years, Tang *et al.* studied corrosion behavior and mechanism of basalt fibers in sodium hydroxide solution.[18] Gutnikov *et al.* researched the correlation of the chemical composition, structure and mechanical properties of basalt continuous fibers.[19]

6.1.1 *Chemical Stability of Basalt Fibers*

The chemical composition of basalt fibers varies from place to place. However, a general characteristic is that it contains a large amount of Al and Fe, hardly appears in glass fibers and has relatively little Ca. Also, not only does a large amount of Fe determine the color of the basalt fiber, but because the atomic weight of Fe is considerable, the density of the basalt fiber is higher than that of the glass fiber.[20] Also, both glass and basalt fibers contain similar amounts of SiO_2. It can be assumed that the composition ratio of SiO_2 and Al_2O_3, and the chemical interaction of Al_2O_3 and -OH groups play a much more significant role in the chemical resistance of basalt fiber.[21] The proportions of the various components that make up the chemical composition of basalt fiber are as follows: $SiO_2 > Al_2O_3 > Fe_2O_3 > CaO > MgO$. In the case of SiO_2, the OH- ions in alkaline solutions break the siloxane bonds and release silicates into the solution, breaking down the fibers. Therefore, it can be assumed that the chemical stability of basalt fibers will be poor under alkaline conditions.[22]

Various studies have revealed this phenomenon. First, it can be seen from the data on the weight change of basalt fiber with time in an alkaline aqueous solution, which is shown in the study of Lee *et al.*[17] and Figs. 6.1 and 6.2.

In Fig. 6.1, basalt fibers are immersed in 30% NH_3, 10% KOH, 10% NaOH and saturated $Ca(OH)_2$ solution for 90 days. Basalt

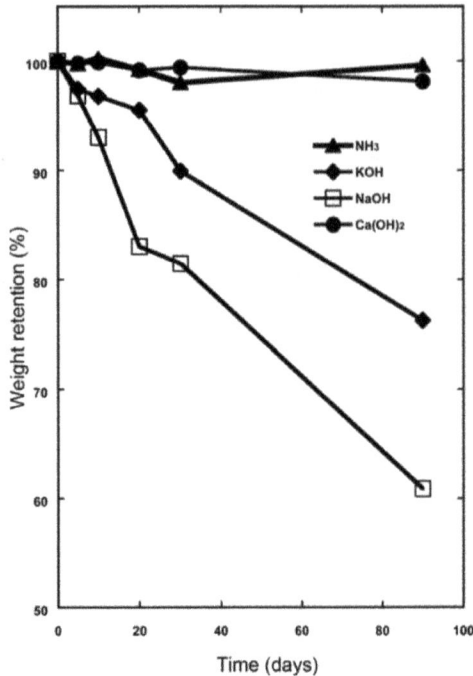

Fig. 6.1 Weight retention of basalt fiber in alkaline solutions (NH_3 30%, KOH 10%, NaOH 10% and saturated $Ca(OH)_2$).[17]

fiber was very stable, with little change in weight retention in 30% NH3 and saturated $Ca(OH)_2$ solution. However, a sharp decrease in weight retention was observed in 10% NaOH solution, and the change continued even in 10% KOH solution. As previously mentioned that the stability of SiO_2 will be weak in influential OH- media, the instability of basalt fibers in NaOH and KOH solutions seems to result from the high dissociation constant. As can be seen in Fig. 6.2, cracks occur due to degradation of the basalt fibers in a 10% NaOH solution. Furthermore, the fragments due to the cracks fall away from the basalt fibers, resulting in a significant decrease in mass.

Tang et al.[18] studied corrosion behavior and mechanism of basalt fibers in sodium hydroxide solution. They measured the mass loss. The mass loss is related to the weighing of the fiber after various treatment periods in the NaOH solution. The mass of basalt

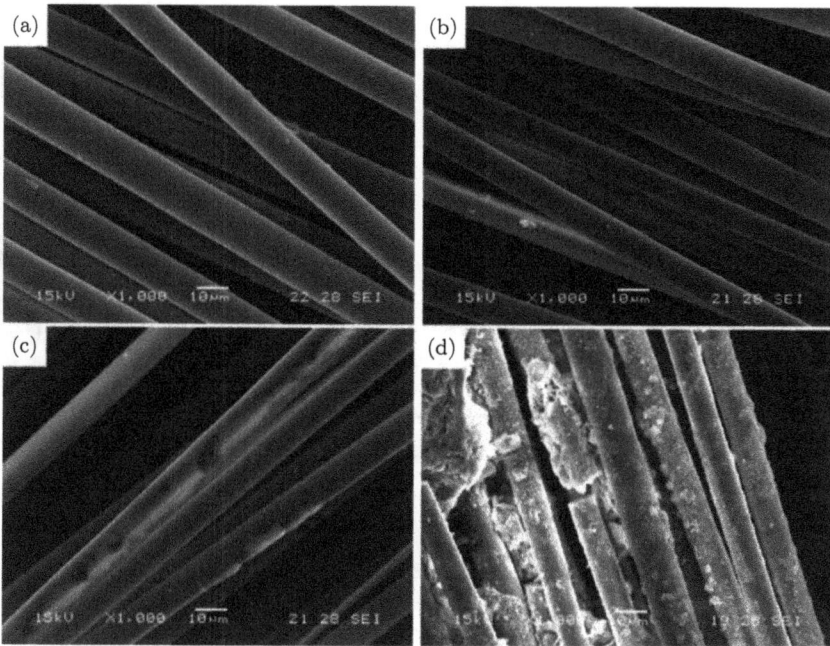

Fig. 6.2 Damaged surface of the fibers after immersion; (a) NH₃ 30% solution, (b) Ca(OH)₂ in saturated Ca(OH)₂ solution, (c) KOH 10% solution and (d) NaOH 10% solution.[17]

fibers before and after corrosion was measured using an electronic analytical balance, and the mass loss of each sample was calculated using Equation (6.1).

$$\text{Mass loss } (\%) = \frac{M0 - M1}{M0} \times 100\% \qquad (6.1)$$

Figures 6.3(a), (b) and (c) show the ratio of the decomposed basalt fiber mass loss to corrosion time, concentration and temperature. Figure 6.3(a) shows that the mass loss rate of basalt fibers is relatively stable with increasing immersion time at a temperature of 25°C. However, at a temperature of 50°C, there was a significant weight loss. More specifically, it is as follows. When the temperature was 25°C, the mass loss rates of fibers with corrosion times of 6 h, 1 day, 2 days and 3 days were 0.3%, 1.2%, 2.2% and 2.4%,

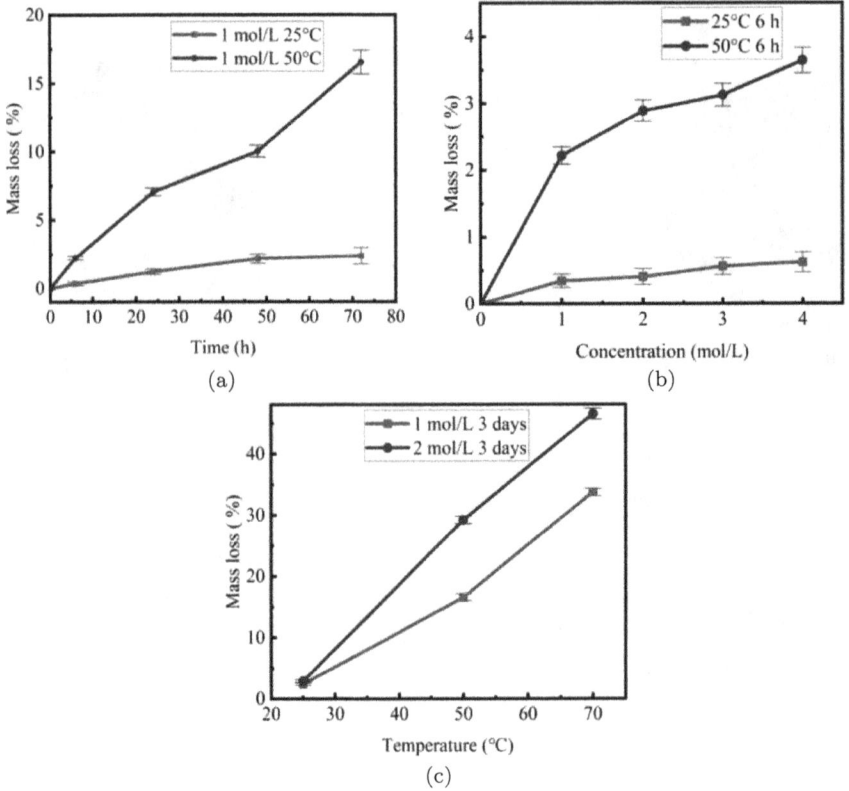

Fig. 6.3 Mass loss ratios of basalt fibers after NaOH solution treatment: (a) mass loss ratios vs. corrosion time; (b) mass loss ratios vs. concentration with 6 h immersion; (c) mass loss ratio vs. temperature with 3 days' immersion.[18]

respectively. When the temperature was 50°C, the mass loss rates of fibers with corrosion times of 6 h, 1 day, 2 days and 3 days were 2.2%, 7.1%, 10.0% and 16.6%, respectively. In Fig. 6.3(b), the mass loss with concentration at the same 6 h shows different mass loss with temperature. The mass loss of the fibers at 50°C was much higher than at 25°C. Figure 6.3(c) shows that when the temperature rises from 25°C to 70°C, the mass loss value of the fiber treated for three days increased sharply.

At a concentration of 1 mol/L and a corrosion time of three days, the mass loss rates of basalt fibers at 25°C, 50°C and 70°C were

Table 6.1 The comparison of XRF results of the desized and the degraded basalt fibers immersed in 1 mol/L NaOH solution at 70°C for three days[18]

Component percentage	SiO_2	Al_2O_3	K_2O	P_2O_5	Fe_2O_3	CaO	MgO	TiO_2	Na_2O	MnO
Desized basalt fibers (wt.%)	47.0	15.1	3.1	0.3	16.0	9.2	4.1	1.4	3.5	0.2
Degraded basalt fibers (wt. %)	33.9	8.1	2.0	0.1	25.0	15.5	8.8	2.4	3.7	0.5
Percentage change (%)	+13.1	+7.0	+1.1	+0.2	−9.0	−6.3	−4.7	−1.0	−0.2	−0.3

2.4%, 16.6% and 33.8%, respectively. The mass loss of basalt fiber increases with increasing temperature, corrosion time and solution concentration. In particular, the mass loss increased with increasing temperature. This proved that higher temperatures accelerate the corrosion of basalt fibers in alkaline solutions.

Tang *et al.*[18] investigated the basalt fibers before and after corrosion by quantitative XRF analysis. The main chemical compositions of desized and degraded fibers are shown in Table 6.1. Compared to desized basalt fibers, the content of SiO_2, Al_2O_3 and K_2O decreased by 13.1%, 7.0% and 1.1%, respectively. The leaching of silicon, aluminum and potassium ions mainly caused this phenomenon.

6.1.2 *Tensile Behavior of Basalt Fibers*

Measure the tensile strength of the basalt fibers immersed in NaOH solution ratio to corrosion time, as shown in Fig. 6.4.[18] At a temperature of 25°C, the tensile strength of the basalt fiber decreased with increasing corrosion time. Tensile strength after 1, 3, 6, 24 and 72 h was 77.9%, 70.7%, 65.4%, 62.5% and 53.6%, respectively. When the temperature was 50°C, the tensile strength decreased and then increased with increasing corrosion time. After 1, 3, 6, 24 and 72 h, the tensile strength was 67.6%, 57.8%, 52.5%, 49.0% and 58.2%, respectively. When the temperature was 70°C, the tensile strength of the decomposed basalt fiber continued to decrease as the corrosion time increased. Tensile strength after 1, 3, 6 and 24 h was 53.5%, 49.2%, 42.0% and 24.6%, respectively.

Many researchers note the effect of SiO_2 content on the mechanical properties of basalt fibers. As can be seen Ref.,[23] not only can SiO_2 reinforce the basalt glass net structure but also form a network and increase the tensile strength of the basalt fiber. Furthermore, the tensile strength of the basalt fiber decreases as the silicon oxide content decreases. This can be confirmed again through the data in Table 6.1 and Fig. 6.4.

The same effect can be expected for Al_2O_3. However, since aluminum atoms can act as both network formers and modifiers, a strong correlation, in this case, is only possible with precise compositional intervals.[24] Increasing the content of CaO and MgO oxides in basalt may increase the ability to crystallize basalt fibers in the base.[25] This significantly reduces the basalt fiber tensile strength. The high concentration of the modifier composition (Na_2O, K_2O, CaO and MgO) reduces the tensile strength of the fiber.[26]

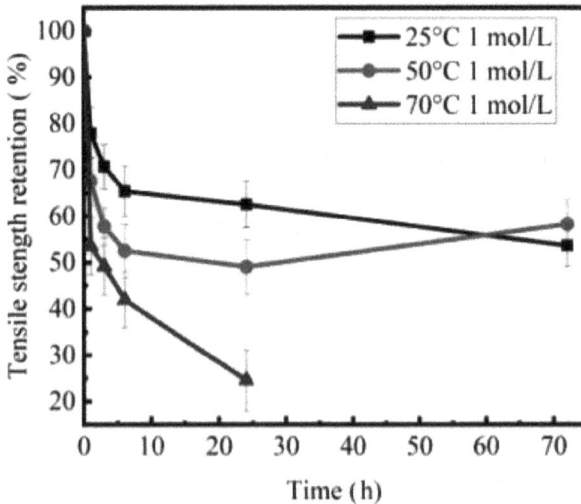

Fig. 6.4 Tensile strength retention ratios of basalt fibers after NaOH solution treatment: tensile strength retention ratios vs. corrosion time.[18]

6.1.3 *Corrosion Process of Basalt Fibers in NaOH Solution*

Studies on the chemical stability of glass or fiber glass in alkali or aqueous solutions have been carried out.[27, 28] Basalt fibers have similar chemical compositions to glass fibers, so there were similarities between the two. Based on the above results and analysis, in Fig. 6.5 Tang *et al.* described the schematic corrosion process of basalt fibers in NaOH solution into three significant corrosion steps: the silicate dissolution step (step 1), the formation and growth of a corrosion shell (step 2) and corrosion peeling (step 3).[18]

In the initial state, it was assumed that the surface of the basalt fiber contained some defects (micro-cracks and pores), but these were so small that they were barely visible. In step 1, the hydroxyl ions in the NaOH solution interfere with the glass fiber — like -Si–O–Si– and -Si–O–Al– bonds in an alkaline environment.[17–28] The aluminosilicate network begins to dissolve through micro-cracks and pores, leading to small pores and micro-cracks that extend to the core of the fiber. As a result, the stress concentration increases, and the tensile strength decreases rapidly. Besides, the hydration reaction was accompanied by the dissolution of basalt fibers, and the insoluble hydroxide formed by calcium and iron covered the fiber surface. A thin hydration layer (corrosive shell) was formed covering

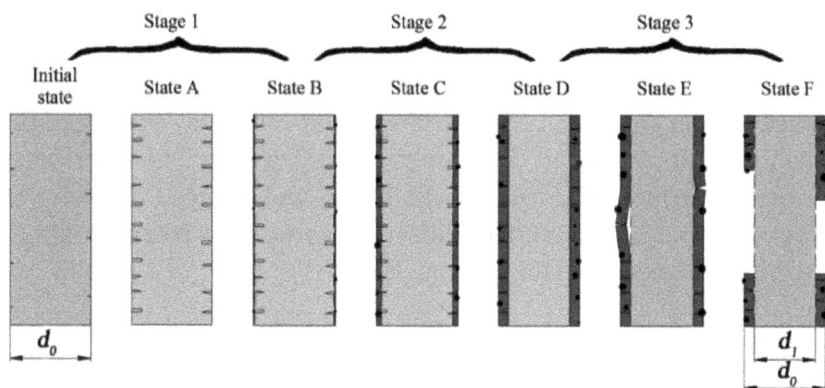

Fig. 6.5 Schematic drawing of the corrosion process of basalt fibers in NaOH solution.[18]

the entire filament surface (state B), and the hydration reaction continued. The corrosion shell acts as a protective layer and slows the dissolution of the fibers.

Passing through state B, the corrosion process proceeded in two stages. As the corrosion time increased, the growth of the corrosion shell gradually slowed the diffusion rate of hydroxyl ions into the fiber core, reducing the dissolution rate of silicate. One end of the holes and micro-cracks was in the corrosive shell and the other end in the fiber core (see state C). The corrosion shells, pores and micro-cracks affect the tensile strength of the basalt fiber, so the basalt fiber had an insufficient tensile strength under this corrosion condition.

In state D, the corrosion shell was the thickest, and almost all holes and micro-cracks were formed in the corrosion shell. The degree of stress concentration is effectively reduced due to the low tensile strength of the corrosion shell and the corrosion shell weakly bonded to the fiber core. The thickness of the corrosion shell mainly influenced the tensile strength of the basalt fiber in this corrosive condition. Compared to the larger fiber diameter, the thickness of the corrosion shell has less effect on the tensile strength reduction than the stress concentration effect on the tensile strength reduction. Thus, the tensile strength of the basalt fiber increased in state D compared to state C.

Water molecules penetrated the glass network,[29, 30] increasing the volume and expansion of the corrosion shell as corrosion proceeded further. Corrosive crusts started to fall off the basalt fiber surface in some areas (state E). In state F, the corrosive shell wholly or partially detaches from the fiber core, the inner (non-corroded) surface of the basalt fiber appears, and the diameter dl is reduced. The inner (non-corrosive) surface of the basalt fiber and the reduced diameter dl increase the tensile strength of the basalt fiber. After passing state D, corrosion proceeded again in state A.

6.1.4 *Conclusion*

In this section, basalt fibers were immersed in various alkaline solutions, and weight retention and tensile strength were observed to find out the chemical stability of basalt fibers in alkaline solutions.

It has been confirmed that weight retention is affected by the type of alkaline solution. However, regardless of the type of alkaline solution used, the tensile strength decreased significantly during the initial stage of immersion.

Also, the weight retention of basalt fiber in 0.4% NaOH solution was much higher than that of a 10% NaOH solution. However, the decrease in tensile strength remained independent of concentration and still fell significantly.

In particular, the corrosion behavior of the basalt fiber according to temperature in NaOH solution showed a different tendency of change in tensile strength as the corrosion time increased. At 70°C high temperature, the corrosion rate of the basalt fiber accelerated, increasing the mass loss rate of the basalt fiber as the temperature increased due to the leaching of silicon, aluminum and potassium ions.

6.2 Tribological Properties of Basalt Fiber Reinforced Composite Materials

Due to many global environmental issues in the past 20 years, researchers have been actively studying for the invention of natural fabrics in the field of fiber reinforced polymer composite materials. For a solution to environmental pollution, usage of the inorganic compound, such as basalt, is being widely suggested. Recently, many in-depth types of researches on basalt materials have been done. In the past 10 years, researchers analyzed the characteristic of basalt fabric, and the continuous and filament fiber basalts, which were used in compound materials.[31, 32] Chemical resistance is one of the essential characteristics of fiber reinforced composite materials. For example, if a crack runs in resin, acid or alkaline solution could permeate through the fiber reinforced composite materials such as fiberglass and basalt fiber, resulting in exposure to corrosion. Nevertheless, no researches or data have come out yet to analyze the various types of damage that occur in fiber reinforced composite materials caused by the harmful amount of aqueous sulfuric acid.

It is essential to collect the data of fundamental substances of a particular material. This study is especially true in the case of composite materials in the beginning stage because an explanation to the aggregate data about the friction wear of basalt fiber reinforced composite material during corrosion is needed. We expect that this experiment can provide further explanation of the corrosion mechanism and heat occurrence when basalt fiber reinforced composite material is applied in a harsh corrosion environment. So, the objective of this study is to investigate the wear friction in identical corrosion environments, mechanism and behavior of wear friction in several different corrosion conditions with basalt fiber reinforced composite material.

The basalt fiber used in the experiment was made out of brick, which is a product of SECOTECH Incorporation. For Vacuum assisted resin transfer molding (VaRTM) formation, epoxy YD-125, a type of thermosetting resin made by Kuk-Do Chemical Company, was used. The thickness of the specimen was 3 mm, with a diameter of 30 mm.

Experimenters exposed all sides of the specimens for surface treatment. Sloan and Symour,[33] and Segovia[34] are some of the well-known experimenters who used the same method. The effectiveness of this method is that it boosts the penetration and corrosion of the sulfuric acid aqueous solution. In other words, it reduces the time of penetration of the solution and thus makes the experiment much easier. The concentration of the sulfuric acid aqueous solution is an essential parameter of this experiment. The variation of concentration of the container holding that solution also affects the corrosion of the specimen. As Carpenter and Kumosa,[35] and Kumosa et al.[36] did, the experimenters measured the concentration ratio before and after the experiment, but they did not in the middle of the procedure. In order to minimize any possible change in concentration, experimenters sealed the uncorroded container with a plastic wrapper before closing it with a cap. Table 6.2 shows whether the specimens were exposed to the sulfuric acid solution, the concentration of the sulfuric acid solution, the exposure time of the sulfuric acid solution and the number of samples.

Swiss CSM's Pin-On Disk friction wear machine was used and the experimental conditions were as follows: radius 3.00 [mm], Lin.

Table 6.2 Test condition of friction-wear specimens

(%)	0 h	150 h	250 h	500 h	720 h
0	3 pcs				
1		3 pcs	3 pcs	3 pcs	3 pcs
5		3 pcs	3 pcs	3 pcs	3 pcs
10		3 pcs	3 pcs	3 pcs	3 pcs
20		3 pcs	3 pcs	3 pcs	3 pcs
35		3 pcs	3 pcs	3 pcs	3 pcs

Speed 15.08 [cm/s], Normal load 10.00 [N], Stop condit. 200.00 [m], Effective Stop: Meters, Acquisition rate: 0.5 [hz], the temperature was maintained at 25°C, and the humidity was maintained at 50%. The friction wear ball used was 6 mm in diameter, and the material was Al_2O_3.

6.2.1 *Weight Change According to the Concentration of Sulfuric Acid Aqueous Solution and Time*

The specimen was removed from the sulfuric acid aqueous solution, washed with distilled water, dried for five days at room temperature. Then the weight change before and after corrosion was measured. It had shown an increase when the H_2SO_4 concentration rate reached 1%, 20% and 35%. When the concentration rate was at 35% H_2SO_4 at 500 h, the increase was significantly higher. At the rate of 5% and 10%, the numbers decreased at 500 h in the beginning stage, but later gradually increased.

The weight increased at 150 h apart from the increase of the concentration rate. However, at 250 h, the weight of the specimen increased in ratio to the increase of the concentration rate. An increase occurred in general, and the experimenters hypothesize that the formation of the precipitate resulting from the chemical reactions between basalt fiber and the sulfuric acid is the leading cause. Figure 6.6 shows the change in weight of the specimen before and after corrosion.

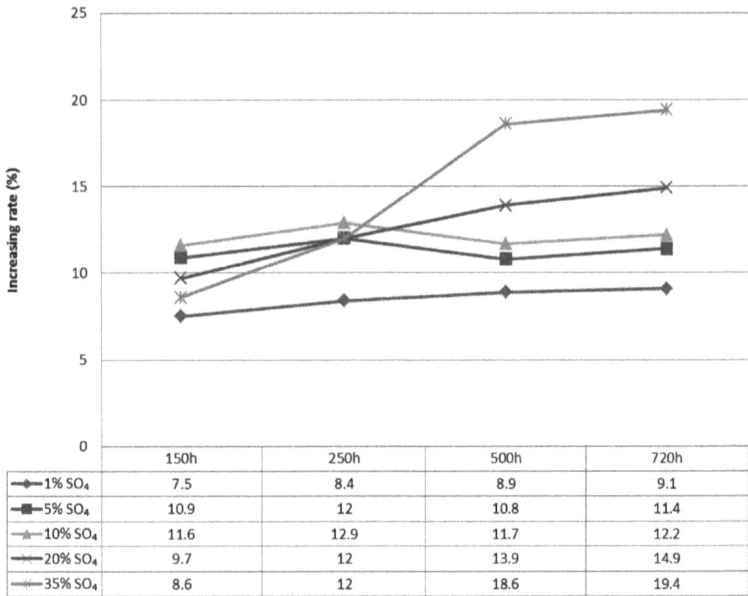

	150h	250h	500h	720h
—◆—1% SO₄	7.5	8.4	8.9	9.1
—■—5% SO₄	10.9	12	10.8	11.4
—▲—10% SO₄	11.6	12.9	11.7	12.2
—✕—20% SO₄	9.7	12	13.9	14.9
—✱—35% SO₄	8.6	12	18.6	19.4

Fig. 6.6   The weight change rate of the specimen before and after corrosion.

6.2.2 Friction and Wear Characteristics
of Specimens before Corrosion

Unlike the wear of metal in a specified pattern, when looking at the shape of the friction wear of the specimen before corrosion, the direction of wear or uniform wear depth has not formed, but an irregular wear shape is shown (Figs. 6.7 and 6.8). This is the reason why the basalt fiber, a reinforcing fiber, was woven in a plain form. Therefore, it can be seen that wear in various shapes occurred according to the angle of weaving. In general, friction wear tests, depending on the material, the wear powder is not blown away, and it is stacked on the root of the wear part, which sometimes affects the friction wear. However, in the case of the basalt fiber reinforced composite material, the phenomenon of lamination of abrasion powder has not been found, and it can be observed that both the basalt fiber and the epoxy resin are worn. The coefficient of friction of the un-corroded specimen is 0.355 and Fig. 6.9 shows a

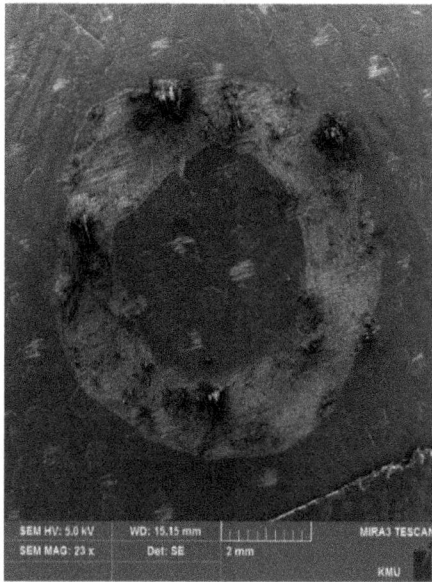

Fig. 6.7 Shape of the friction and wear of the intact specimen.

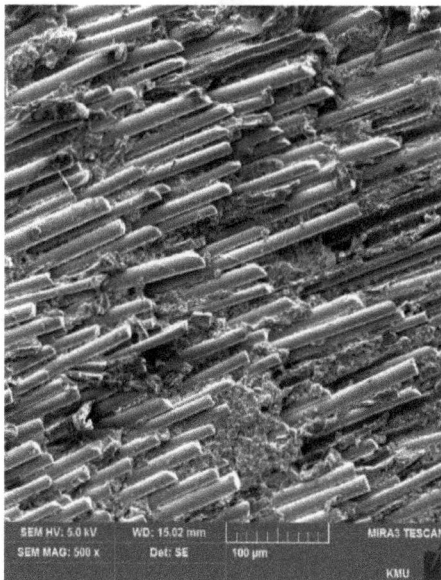

Fig. 6.8 SEM of the intact specimen after friction and wear test.

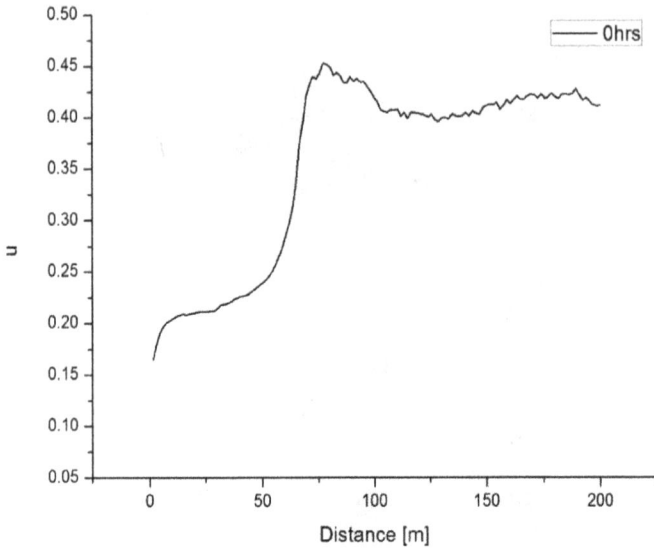

Fig. 6.9   Friction coefficient graph of an intact specimen.

graph of the coefficient of friction against the wear distance during frictional wear. The friction wear distance where the coefficient of friction in the graph reaches the maximum is 77.18 m, and the coefficient of friction is 0.45. When calculated using the wear area, it is about 0.03 mm from the surface, and the total thickness of the specimen is 3.35 mm. These results can be assumed to be where the fiber begins to cross the surface of the epoxy resin (Fig. 6.10).

6.2.3   *Frictional Wear Characteristics According to the Concentration of Sulfuric Acid Aqueous Solution at 150 h*

For 0% sulfuric acid aqueous solution, the tribometer was 0.36, 0.16 for 1%, 0.13 for 5%, 0.17 for 10%, 0.15 for 20% and 0.11 for 35%. In general, when the coefficient of friction and wear shows a deviation of 0.05, there is no change in the coefficient of friction. Figure 6.11 shows the same coefficient of frictional wear from the surface of the corroded specimen to the inside. It is thought that the coefficient

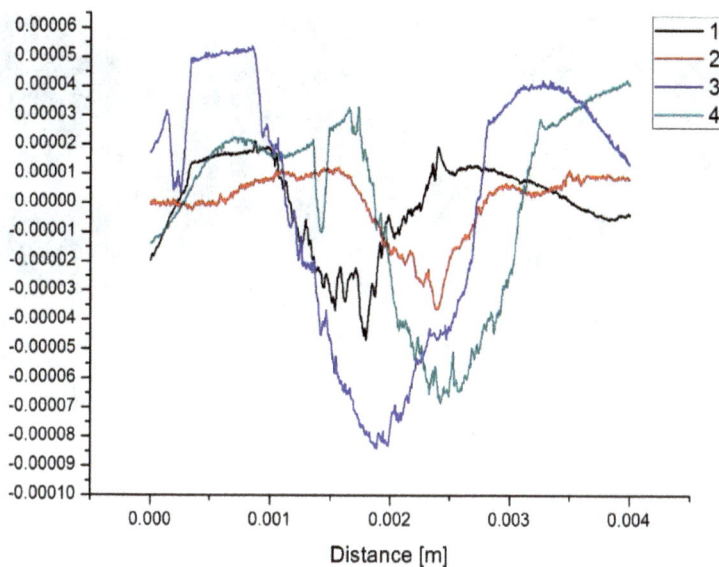

Fig. 6.10   Wear area graph of the intact specimen with four areas.

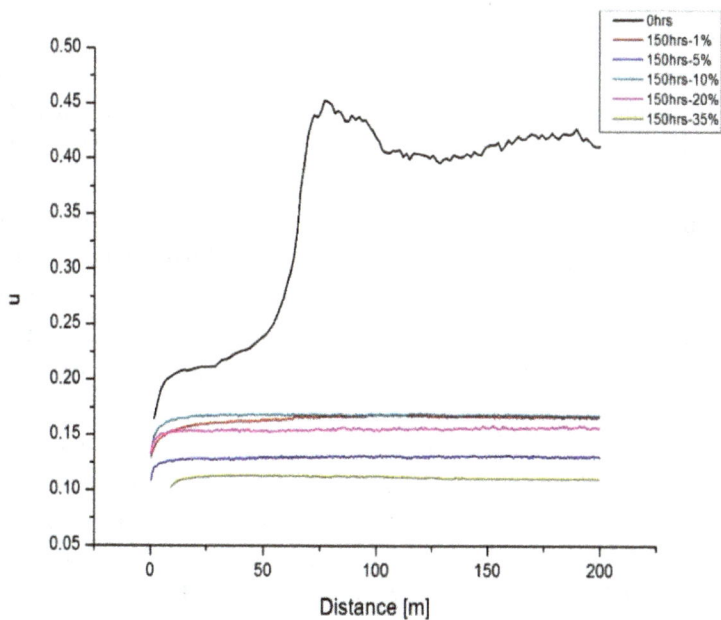

Fig. 6.11   Wear area graph of the intact specimen with four areas.

J.-C. Ha et al.

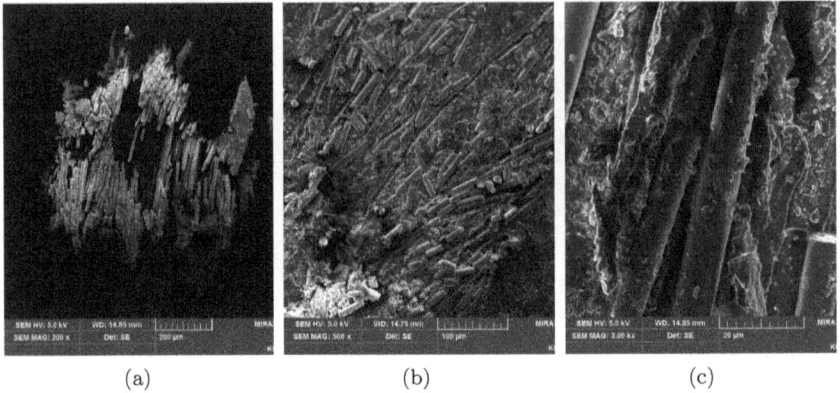

(a)   (b)   (c)

Fig. 6.12   SEM of 10% sulfuric acid with 150 h.

of friction was lowered due to the chemical reaction caused by the sulfuric acid aqueous solution on the surface of the specimen.

We observed the specimens subjected to friction wear at 10% and 35% concentration through SEM. In the case of Fig. 6.12(a), it can be seen that the frictional wear does not proceed while forming all of one circular trajectory. Epoxy progresses in a form that breaks while forming a lump. This is due to the chemical reaction between the epoxy and the sulfuric acid aqueous solution. In Fig. 6.12(b), frictional wear has progressed in the part where the basalt fiber and epoxy are mixed inside the specimen. It can be seen that the worn basalt fiber and epoxy do not fall off in the form of dust, but the worn part remains in the form of dust, which affects the frictional wear. In Fig. 6.12(c), the ball pressure during frictional wear made a shape in which epoxy solidifies and precipitates when corrosion is in progress on the fiber.

Specimens exposed to a 35% sulfuric acid aqueous solution showing a relatively low coefficient of friction were examined. In Fig. 6.13(a), the peeling of resin surface was lifted, and voids were formed between the surface resin and the bonded fibers by a chemical reaction by the sulfuric acid aqueous solution so that the surface resin was lifted in the form of a dome. In the case of Fig. 6.13(b), the order

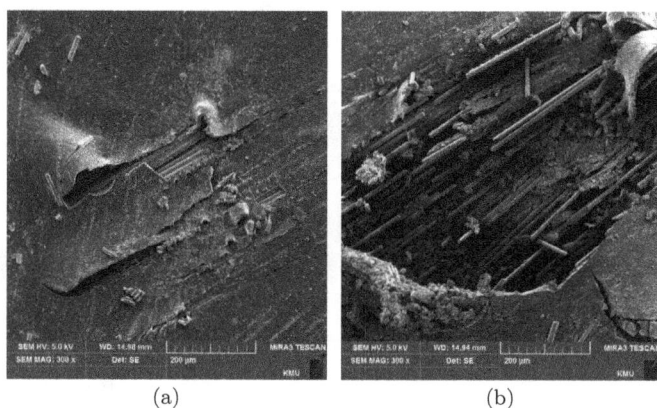

(a)   (b)

Fig. 6.13 SEM of 35% sulfuric acid with 150 h.

of frictional wear can be seen. As continuous friction wear progresses, wear occurs sequentially in the order of resin, fiber and resin.

6.2.4 *Friction and Wear Characteristics According to the Concentration of Sulfuric Acid Aqueous Solution at 250 h*

Figure 6.14 shows in 0% sulfuric acid aqueous solution, the tribometer is 0.36, 0.11 for 1%, 0.19 for 5%, 0.10 for 10%, 0.12 for 20% and 0.17 for 35%. It can be seen that there is no significant difference in the friction coefficient deviation according to the concentration change. Furthermore, the graph of the coefficient of friction shows that the same phenomenon as the specimen corroded in 150 h.

Specimens exposed to 20% sulfuric acid aqueous solution were observed with SEM. In Fig. 6.15(a), the surface resin and the basalt fiber wear in each direction due to frictional wear on the perforated area by the sulfuric acid aqueous solution can be seen. In Fig. 6.15(b), it can be seen that the surface resin remains within the circular trajectory of friction wear. It can be seen that it is not a general concept that the fibers are worn after all the surface resins are worn, but are made together.

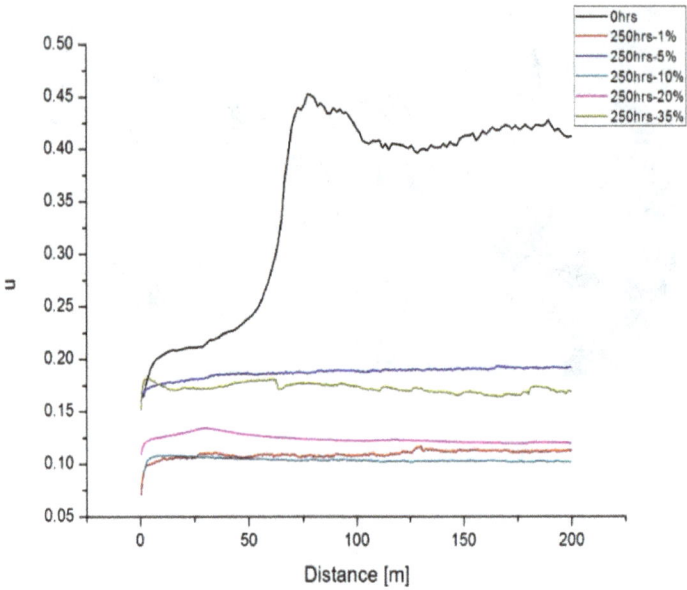

Fig. 6.14 Friction coefficient graph of 250 h with several sulfuric acid percentages.

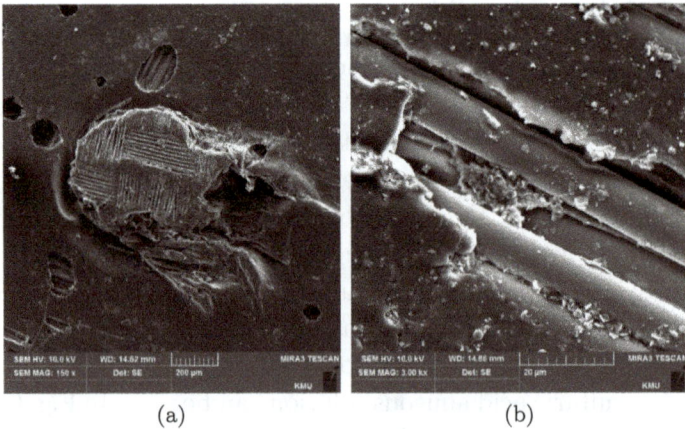

(a)   (b)

Fig. 6.15 Friction and wear shape (a) and non-friction and wear surface (b) of 20% sulfuric acid with 250 h.

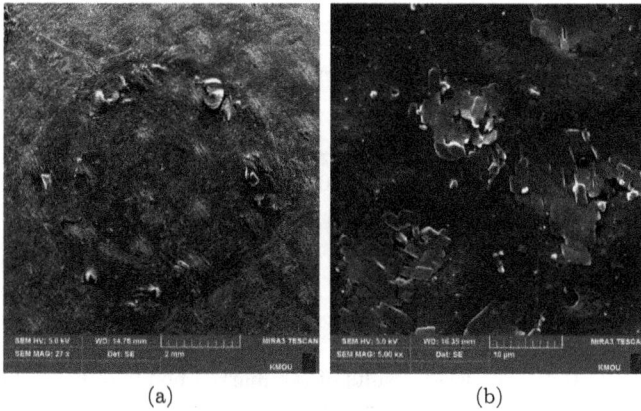

(a)   (b)

Fig. 6.16   Friction and wear shape (a) and non-friction and wear surface (b) of 35% sulfuric acid with 250 h.

Figure 6.16(a) is a specimen corroded in a 35% sulfuric acid aqueous solution, and when examining the trajectory of frictional wear, it can be seen that the trace is not clear. This means that there is not much wear due to the low coefficient of friction. Figure 6.16(b) is the inner surface of the circular orbit that has undergone frictional wear, and it can be observed that the resin has already been broken into blocks by the sulfuric acid aqueous solution and cracks have occurred. When friction wear progresses on this part, it can be observed in Fig. 6.16(a) that the resin quickly wears out. EDX tests were performed at the four marked locations to observe what chemical components were on the specimen surface. In Table 6.3, it can be seen that the sum of the components of the two elements is 85% left and right, with 45% left and right of element C and 40% left and right of element O on average in all four locations. It can be seen that the two elements are the cause of lowering the coefficient of friction.

The EDX experiment was performed on the basalt fiber inside the 35% test piece as shown in Table 6.4. The chemical composition is almost the same regardless of whether or not precipitates are formed. In particular, it can be seen that the sum of the contents of C and O elements range from 53% to 70%. This is similar in comparison

Table 6.3   The EDX results on the specimen surface

Element content (wt%)								
C	N	O	Si	Mo	Co	Na	Pt	S
a	43.65	07.20	39.98		09.16			06.56	04.46
b	47.97	06.10	34.92						
c	53.13		34.78	01.06				06.79	04.24
d	43.12	06.18	27.66			03.68	01.53	10.96	06.87

Table 6.4   The EDX results of the marked points of Fig. 6.14

Element content (w%)									
C	O	Fe	Na	Mg	Al	Si	S	Mo	Pt
a	16.75	44.13	15.28	01.15	01.37	04.75	14.15	02.43		
b	31.07	38.28	10.33	01.03	00.78	03.29	12.12		03.09	
c	13.22	40.76	12.41	02.35	01.83	05.72	17.47			06.24

with the content of the corroded specimen surface. The reason that the coefficient of friction between the surface and the interior of the test piece remains similar can be assumed to be that the sum of the contents of C and O elements in the surface and fiber are similar. Comparing with the results of the EDX test on the surface of the specimen, it can be seen that Fe, Mg and Al elements were detected only in the basalt fiber.

6.2.5   *Friction and Wear Characteristics According to the Concentration of Sulfuric Acid Aqueous Solution at 500 h*

For 0% sulfuric acid aqueous solution, the tribometer was 0.36, 0.14 for 1%, 0.12 for 5%, 0.14 for 10%, 0.14 for 20% and 0.17 for 35% (Fig. 6.17). It can be seen that there is no significant difference in the friction coefficient deviation according to the concentration change. The graph of the coefficient of friction shows that the same phenomenon as the specimen corroded in 250 h (Fig. 6.18).

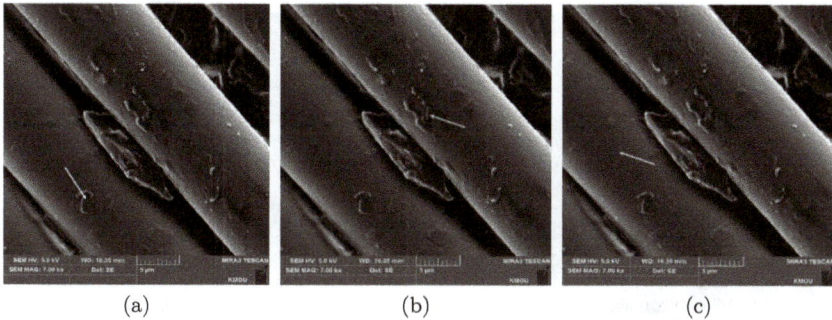

(a)   (b)   (c)

Fig. 6.17 SEM of friction and wear basalt fiber of 35% sulfuric acid at 500 h.

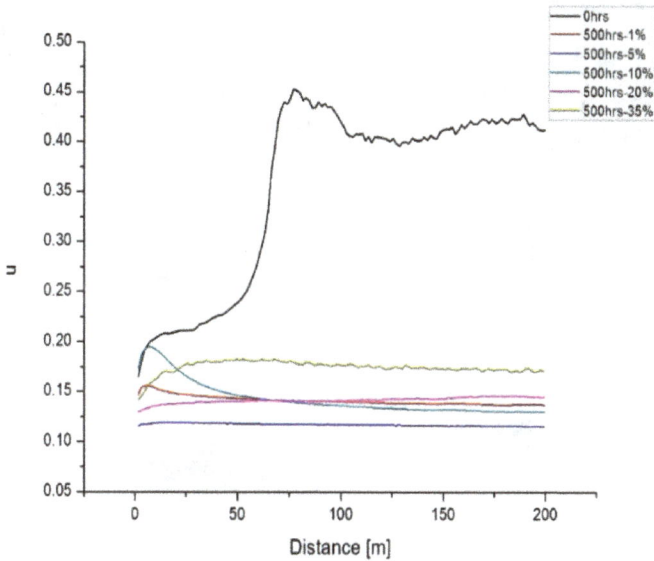

Fig. 6.18 Friction coefficient graph of 500 h with several sulfuric acid percentages.

Looking at the specimen under a microscope, the shape raised on the specimen surface and the epoxy surface disappeared due to friction, and the basalt fibers can be observed. Frictional wear progressed under these conditions. This also indicates that similar coefficients of friction are maintained for both the surface protrusion

1% (100x magnification)   5% (100x magnification)

10% (100x magnification)   20% (100x magnification)

35% (100x magnification)

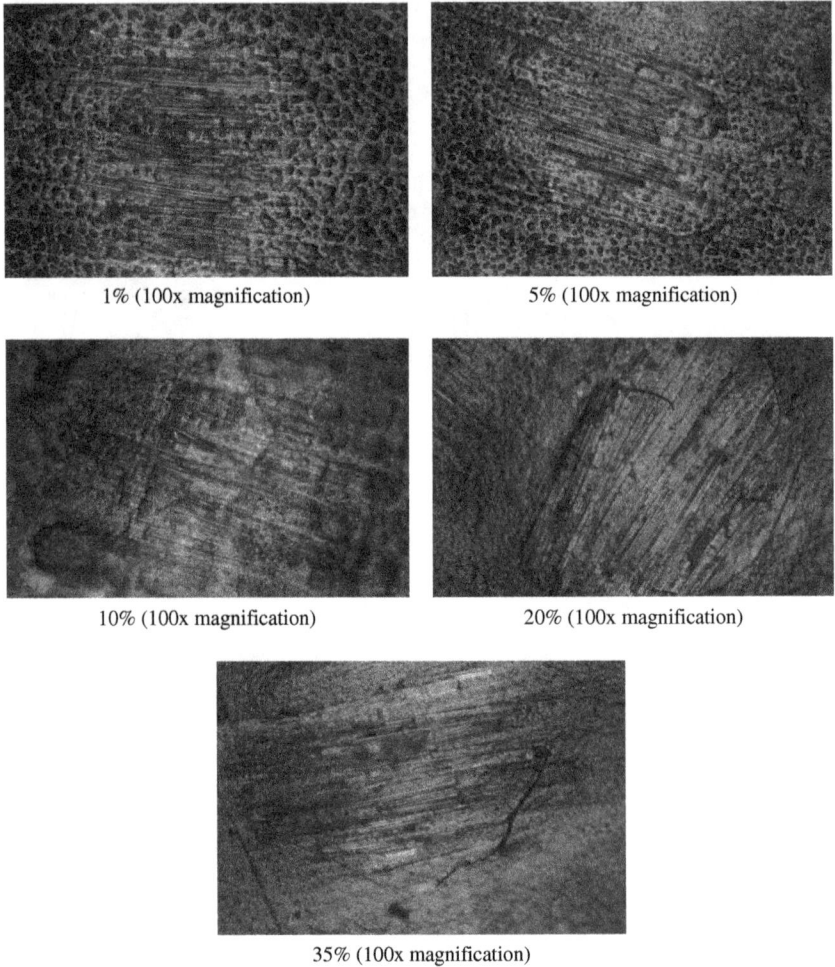

Fig. 6.19 Friction and wear shape by optical microscope at 500 h with several sulfuric acid percentages.

and the basalt fiber. Therefore, it can be judged that the chemical reaction products are similar on the surface of the specimen and the surface of the basalt fiber. Figure 6.19 shows the friction and wear shape by an optical microscope at 500 h with several sulfuric acid percentages.

6.2.6 Friction and Wear Characteristics According to the Concentration of Sulfuric Acid Aqueous Solution at 720 h

For 0% sulfuric acid aqueous solution, the tribometer was 0.36, 0.14 for 1%, 0.20 for 5%, 0.10 for 10%, 0.17 for 20% and 0.12 for 35%. The highest coefficient of friction is 0.20 at 5%, and the lowest coefficient of friction is 0.10 at 10%. In Fig. 6.20, it can be seen that the coefficient of friction continues to increase at 5%, and the coefficient of friction continues to decrease at 35%.

The harshest environment in the friction wear test is shown in Fig. 6.21 where the SEM observation picture of the specimen was exposed to the most extended 720 h, in 35% sulfuric acid aqueous solution. Figure 6.21(a) is a specimen corroded in 35% sulfuric acid aqueous solution, and when examining the trajectory of frictional wear, it can be seen that the trace is not clear. This means that there

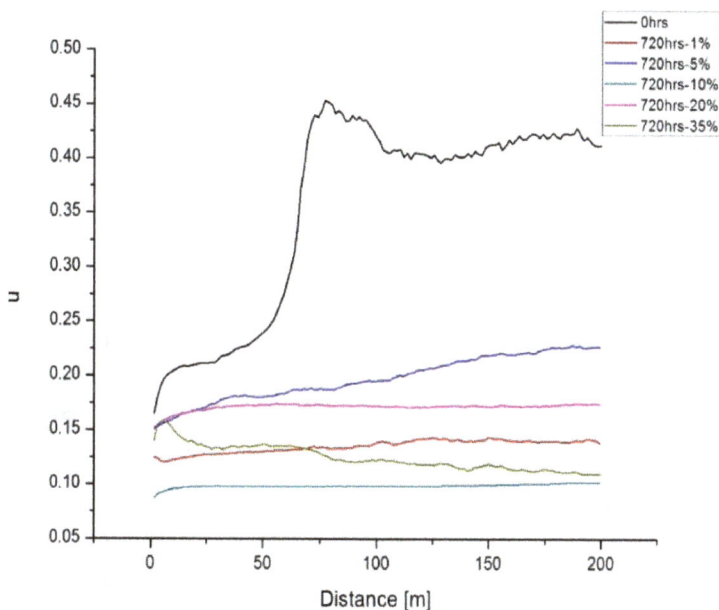

Fig. 6.20 Friction coefficient graph of 720 h with several sulfuric acid percentages.

(a)   (b)

Fig. 6.21 Friction and wear shape (a) and non-friction and wear surface (b) of 35% sulfuric acid at 720 h.

is not much wear due to the low coefficient of friction. Figure 6.21(b) is the inner surface of the circular orbit that has undergone frictional wear, and it can be observed that the resin has already been broken into blocks by the sulfuric acid aqueous solution and cracks have occurred. Compared with the 250 h, 35% sulfuric acid aqueous solution test piece, this resulted in more severe corrosion shape and more precipitates.

The EDX test was performed at three marked locations to observe what chemical components were on the specimen surface. Table 6.5 shows, on average, the amounts of elements C and O occupy from 60% to 92% in all three inspection areas. In the case of No. 2 point, it can be seen that the epoxy resin is oxidized, and the Fe element is precipitated from the basalt fiber inside to form a product.

EDX experiments were performed on the basalt fiber. The amount of chemical composition in Fig. 6.20(a) at the point where the precipitate was formed was compared to the three places b, c, d were it was not. In all four observed points, the sum of C and O elements was distributed from about 60% to 91%. In particular, In Fig. 6.23, (b) and (c) points account for about 90%. In the case of (a) point of the fiber, it was eroded. Comparing with the results of the EDX test on the surface of the specimen, it can also be seen that

Table 6.5 The EDX results of the marked points of Fig. 6.22

	C	O	Fe	Na	Pt	S
1	17.23	47.05		12.44	07.10	16.17
2	14.84	44.33	08.07	14.66	05.02	13.08
3	70.12	22.29				07.59

Fig. 6.22 SEM of non-friction and wear surface of 35% sulfuric acid at 720 h.

Fe, Mg and Al elements were detected only in the basalt fiber equal to 250 h (Table 6.6).

6.2.7 *Conclusions*

When the weight of the specimens before and after corrosion was measured, the weight of all specimens increased. It can be seen that

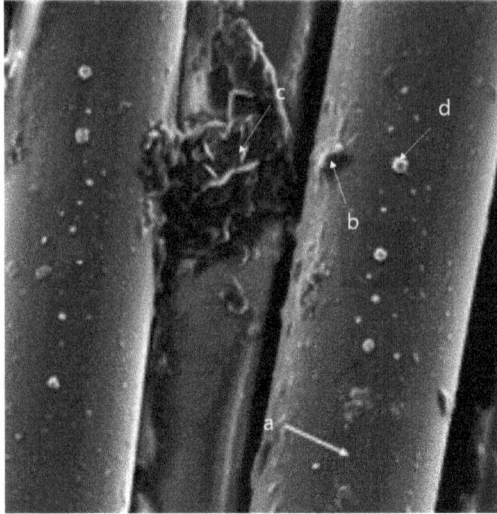

Fig. 6.23 SEM of friction and wear surface of 35% sulfuric acid at 720 h.

Table 6.6 The EDX results of the marked points of Fig. 6.23

	Element content (wt%)								
	C	O	Fe	Na	Mg	Al	Si	S	Zn
a	22.19	38.11	15.55	01.88	01.47	05.87	14.94		
b	65.25	27.37							01.96
c	40.47	45.08		03.85		01.59	01.24	07.76	
d	14.62	42.37	14.07	03.43	01.26	04.78	19.46		

a chemical reaction generated a precipitate. These precipitates were generated both on the surface and inside.

The coefficient of friction of the specimen before corrosion is the highest at 0.36. This is a coefficient of friction that is about 2–3 times higher than that of the specimens after corrosion. After corrosion, the coefficient of friction of the specimen was distributed between 0.1 and 0.2.

Looking at the progress of the friction coefficient of the specimen before corrosion, if the friction coefficient becomes constant after

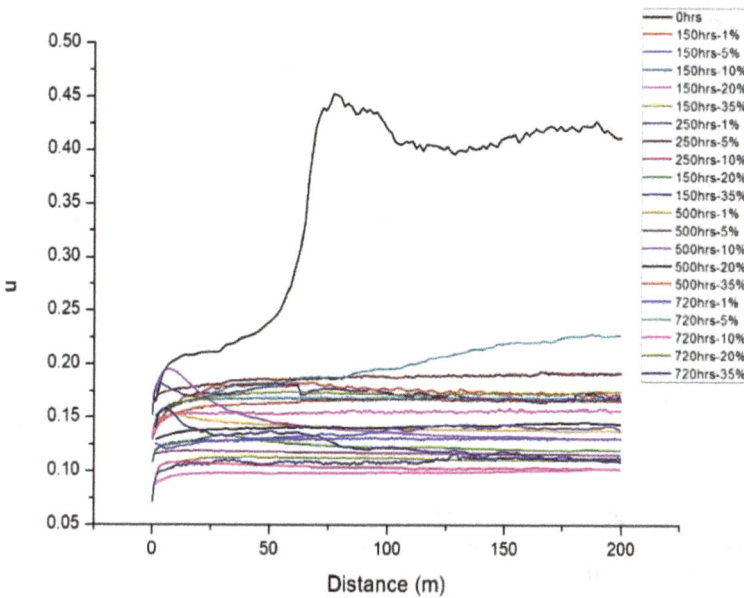

Fig. 6.24 Friction coefficient graph of all specimens.

about 77.19 m, in the case of the specimen after corrosion, the longest distance is about 16.63 m, and the shortest distance is 3.02 m, the friction coefficient becomes constant (Fig. 6.24).

There are two reasons why the coefficient of friction of the specimens after corrosion is lower than that of the specimens before corrosion. First, during the corrosion test in sulfuric acid aqueous solution, the oxide in the basalt fiber causes sagging (a phenomenon that makes the specimen like rubber) of the basalt fiber due to elution. Second, microscopic spaces were generated due to the dissolution of the resin between the basalt fibers due to the sulfuric acid aqueous solution. This space acts with elasticity during the frictional wear test, which is the cause of the physical phenomenon that the frictional force is significantly reduced.

This physical phenomenon significantly reduces the rate of wear. The difference in friction coefficient is under 0.05, explaining that there is no significant difference among time, concentration rate and friction wear when the acid solution is applied.

References

[1] Y. Shan and K. Liao, *Compos: Part B*, 32, 2001, 355–363.
[2] T. K. Tsotsis, S. Keller, K. Kyejune, J. Bardisc and J. Bishd, *Compos. Sci. Technol.*, 61, 2001, 75–86.
[3] P. Mertiny and K. Ursinus, *Polym. Test*, 26, 2007, 751–760.
[4] A. Belarbi and S.-W. Bae, *Compos: Part B*, 38, 2007, 674–684.
[5] A. K. Amiruddin, S. M. Sapuan and A. A. Jaafarb, *Mater. Des.*, 67, 2007, 2643–2654.
[6] T. J. Myers, H. K. Kytömaaa and T. R. Smith *J. Hazard Mater.*, 142, 2007, 695–704.
[7] E. C. Edge, *Composites*, 11, 1980, 101–104.
[8] J. M. Shultz and C. Lhymn, *Polym. Compos.*, 7, 1984, 208–214.
[9] J. N. Price and D. Hull, *Compos. Technol.*, 28, 1987, 193–200.
[10] L. Kumosa, D. Armentrout and M. Kumosa, *Compos. Sci. Technol.*, 61, 2001, 615–623.
[11] A. Akdemir, N. Tarakcioglu and A. Avci, *Compos: Part B*, 32, 2001, 123–129.
[12] E. L. Rodriguez, *J. Mater. Sci. Lett.*, 6, 1987, 718–720.
[13] B. Dewimille and A. R. Bunsell, *Composites*, 14, 1983, 35–40.
[14] G. Huang, *Mater. Des.*, 2008.
[15] C. Scheffler, T. Förster, E. Mäder, G. Heinrich, S. Hempel and V. Mechtcherine, *J. Non-Cryst. Solid.*, 355, 2009, 2588–2595.
[16] B. Wei, H. Cao and S. Song, *Mater. Des.*, 31, 2010, 4244–4250.
[17] J. Lee, J. Song and H. Kim, *Fibers Polym.*, 15, 2014, 2329–2334.
[18] C. Tang, H. Jiang, X. Zhang, G. Li and J. Cui, *Mater. (Basel)*, 11(8), 2018, 1381.
[19] S. Gutnikov, E. Zhukovskaya, S. Popov and B. Lazoryak, *AIMS Mater. Sci.*, 6, 2019, 806–832.
[20] M. Wang, Z. Zhang, Y. Li, M. Li and Z. Sun, *J. Reinfor. Plast. Compos.*, 27, 2008, 393.
[21] Y. Ma, B. Zhu and M. Tan, *Cement Concrete Res.*, 35, 2005, 296.
[22] P. Purnell and J. Beddows, *Cement Concrete Compos.*, 27, 2005, 875.
[23] X. Chen, Y. Zhang and H. Huo, *J. Nat. Fibers*, 2018, 1–9.
[24] S. I. Gutnikov, A. P. Malakho and B. I. Lazoryak, *Russ. J. Inorg. Chem.*, 54, 2009, 191–196.
[25] N. N. Morozov, V. S. Bakunov and E. N. Morozov, *Glass Ceram.*, 58, 2001, 100–104.
[26] K. L. Kuzmin, E. S. Zhukovskaya and S. I. Gutnikov, *Int. J. Appl. Glass Sci.*, 7, 2016, 118–127.
[27] C. Jantzen, K. Brown and J. B. Pickett, *Int. J. Appl. Glass Sci.*, 1, 2010, 38–62.
[28] A. Paul, *J. Mater. Sci.*, 12, 1977, 2246–2268.
[29] H. Scholze, D. Helmreich and I. Bakardjiev, *Glastech. Ber.*, 48, 1975, 237–247.

[30] M. Y. Liu, H. G. Zhu, N. A. Siddiqui, C. K. Y. Leung and J. K. Kim, *Compos. Part A*, 42, 2011, 2051–2059.
[31] J. Sim, C. Park and D. Y. Moon, *Compos. Part B*, 36, 2005, 504–512.
[32] S. E. Artemenko, *Fibre Chem.*, 35(3), 2003, 226–229.
[33] F. E. Sloan and R. J. Seymour, *J. Compos. Mater.*, 26(35), 1992, 2655.
[34] F. Segovia, M. D. Salvadore, O. Sahquillo and A. Vicente, *J. Compos. Mater.*, 41, 2007, 16.
[35] S. H. Carpenter and M. Kumosa, *J. Mater. Sci.*, 35, 2000, 4465–4476.
[36] L. Kumosa, A. Armentrout and M. Kumosa, *Compos. Sci. Technol.*, 62, 2002, 1999–2015.

Part 4
In situ Characterization and Applications

Chapter 7

Application of Composite in Aerospace Structure

Do-Hoon Shin

Aerostructure Development Engineering,
Korean Air Aero-Space Division,
Busan, Republic of Korea
dohshin@koreanair.com

Over the past several decades, the autoclave process has remained one of the most robust and stable operations in fabricating structural composite parts for the aerospace industry. Recently, there has been an explosive increase in the use of carbon fiber reinforced composite in the aerospace industry; however, there are some disadvantages in terms of capital investment and operations. In order to overcome the negative characteristics of the autoclave process, a lot of research into manufacturing methods with lower costs, higher production rates and improved processing efficiency has been performed. In this chapter, the autoclave process and the representative Out of Autoclave (OOA) process, which are used mainly in the aerospace industry, will be introduced. Also, some examples used to improve internal defects of aerospace parts will be shown.

7.1 Introduction

To improve aircraft fuel efficiency, the industry has performed a tremendous amount of research. The light weight of aircraft is a very crucial factor in increasing the payload and flying time. As shown in Fig. 7.1, the application rate and demand for fiber reinforced composite in the commercial aircraft has soared since 2000.[1] Various researches on manufacturing methods to overcome cost and the limitation of shapes are being carried out. The conventional manufacturing method for composite parts is standardized with an autoclave

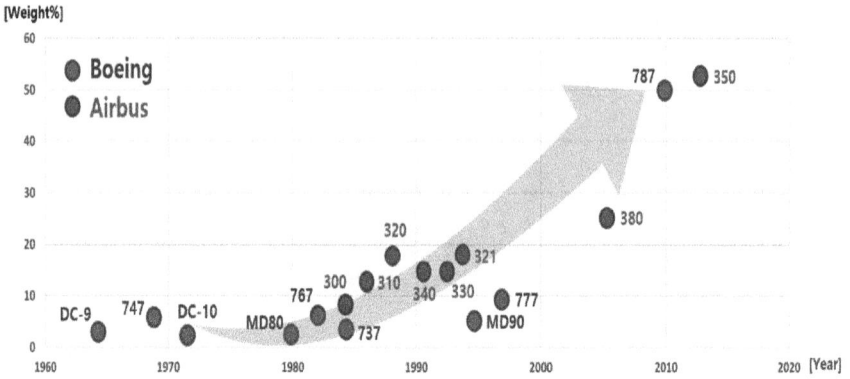

Fig. 7.1   Trend of application of carbon fiber reinforced composite in commercial airplane.

method, which provides high mechanical property and dimensional stability. However, it has a long manufacturing time (6–12 h) and has a limit in the scope of application due to high initial investment cost, labor-intensive manufacturing environment and the technical difficulty in fabricating complex-shaped parts.[2] In order to get over this limitation and to secure more efficient manufacturing methods, research of Out of Autoclave (OOA) process is in progress.[3]

OoA process applies the curing method such as oven or mold heating without autoclave. The manufacturing processes include vacuum bag only (VBO), resin transfer molding (RTM) for thermoset material and thermoforming process for thermoplastic material. Considering the shape and size of parts and production rate, the best manufacturing process can be determined. Although production of aircraft parts using OOA process has been developed mainly on the aircraft secondary structure, an improvement of the process and material is underway in the application of the aircraft primary structure which supports the main loads such as wing and fuselage.

7.2 Autoclave Process

An autoclave is a pressurized vessel that applies heat and pressure simultaneously. The autoclave process is the most widely used method of producing high-quality laminates in the aerospace industry. This curing method has several shortcomings such as the need for a lot of process materials, long process lead time and high initial investment cost. On the other hand, this method enables to maintain vacuum inside while applying extra pressure to the outside bag, which can prevent the entrapped volatiles from growing into void defects.

7.2.1 *The Cure Process*

The autoclave process needs to seal the laminate in a plastic bag after ply collection.

A typical bagging scheme is shown in Fig. 7.2 and each role of process materials which makes up the bagging and the sequence of lay-up is as follows:

(1) Clean the mold or the base plate with solvent acetone and apply mold release on a tool surface to make it easy to detach the cured part.

(2) Place a non-porous release on the mold and a peel ply may be applied directly to the laminate surface, if necessary. At this time, make sure that the non-porous release completely adheres to the prepreg so that there is no artificial damage after curing.

(3) To prevent resin from escaping from the edge of the laminate, dams are placed around the periphery of the lay-up. The dam

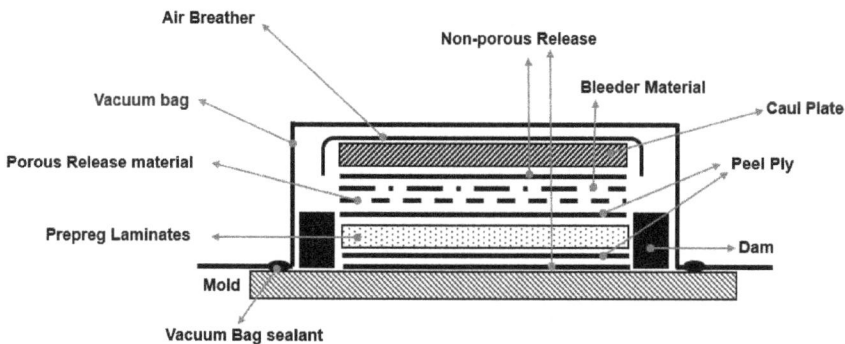

Fig. 7.2 Schematic representation of typical bagging.

should be butted up against the edge of the lay-up to prevent pools from forming between the laminate and dams.

(4) A layer of porous release material is placed over the lay-up. The smaller the hole on the film, the better and the more uniform it needs to be.

(5) The bleeder is placed on the lay-up, since the amount of the bleeder materials can control the content of the fiber and resin content ratio of the composite part. The bleeder should completely cover the lay-up to avoid non-uniformity of the resin content ratio after curing the part.

(6) Place the caul plate on the non-porous film of the lay-up to ensure that the pressure in the autoclave is uniformly transferred to the lay-up.

(7) The lay-up is covered with the air breather to facilitate vacuum removal. The purpose of the breather is to allow air and volatiles to evacuate out of the lay-up during curing.

(8) The vacuum bag is placed on the top of the laminate and it should be sealed with the sealant between the mold and the vacuum bag to prevent the leakage of vacuum.

7.2.2 Cure Cycle

Carbon fiber reinforced composite materials based on thermoset resin are supposed to be heated under pressure during the cure process; the heat promotes the chemical reaction of the prepreg to harden the resin and the pressure absorbs the excess resin to the bleeder and shapes the laminate to a suitable thickness, minimizing the rate of voids.

Therefore, it is important to know how much and for how long to apply the heat and pressure during the cure process. In producing composite materials, the display of the temperature and pressure as the function of time is called the cure cycle and, this cycle depends on the character of the resin. The degree of heating and the duration of applying the heating affect the temperature, viscosity and the rate of cure of the prepreg. The temperature of the prepreg affects the reaction rate of the resin and the amount of reaction within

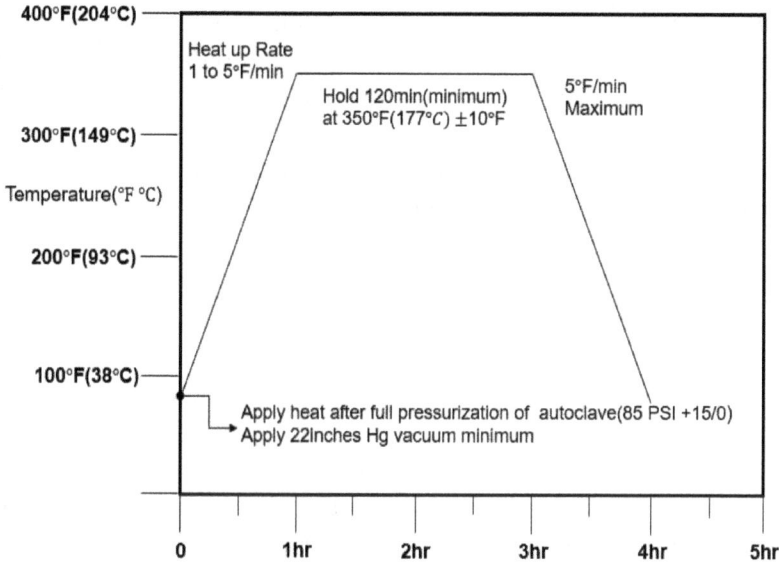

Fig. 7.3 Schematic representation of a typical laminate cure cycle.[4]

a certain time. The degree of the gelation can be evaluated with viscosity change, so the time to apply pressure and vacuum can be adjusted. When the prepreg resin undergoes the chemical changes though exothermic reaction, the degree of the cure can be detected by measuring the calorific value.

The pressure which is introduced while curing has an effect on the resin flow, the thickness of prepreg and the rate of void, etc. It is very important to evaluate the amount of applied pressure, start point and the duration time. Figure 7.3 shows a typical 350°F cure cycle for parts in aerospace industry.

7.2.3 Case Study of Solving Void Issue of a Composite Part Fabricated by Autoclave Process[5]

Void is one of the most stringent requirements which all participants in composite fabrication of aircraft structure must meet. In the aerospace sector, engineering wants to keep voids as low as 1–2%. It is known that void level has detrimental impacts on the mechanical properties of composite structures. Air evacuation and resin pressurization are strongly recommended to control void issue in composite parts.[6-8]

Most of the trapped air in the composite part has been evacuated through the debulk process; however, it is not easy to perfectly remove air from the laminate and even harder for complex-shaped or large-sized parts with surface film on the outer layer. The surface film sometimes disrupts these general solutions to the complex-shaped or larger parts. The objective of this research is to verify how the surface film works negatively in void control during the curing process and how to solve it for the parts with surface films used in aerospace.

Two test articles were fabricated with two different conditions in which surface films were applied fully and partially. The specific materials used for these specimens are CFRP (Cycom 970/T-300, solvay) and surface film material (FM1515, Solvay). In Fig. 7.4, defects on the surface were not inspected, however, from the destructive test some indications were observed on the corner area of the specimens with full surface film. On the other hand, as shown in Fig. 7.5, the other specimen covered partially with surface film except for the corner area showed interesting results contrary to the previous effort. There were no issues except for some minor pin holes on the surface of the corner area. The result of destructive test showed that there were no internal voids or porosities in the laminate, even in the corner area.

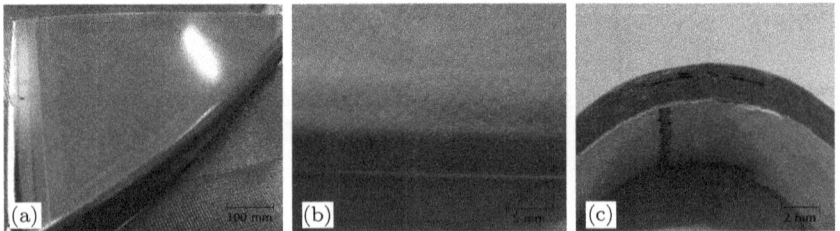

Fig. 7.4 Cured test article (Commercial Aircraft Fairing structure) with full application of surface film.

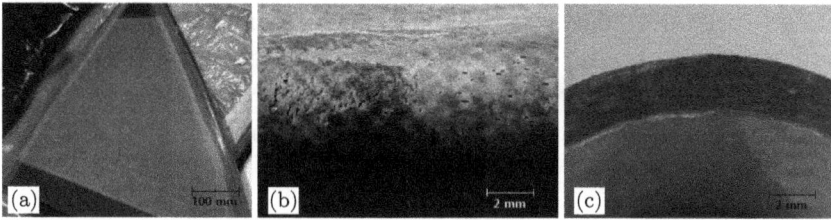

Fig. 7.5 Cured test article with partial application of surface film (with permission).

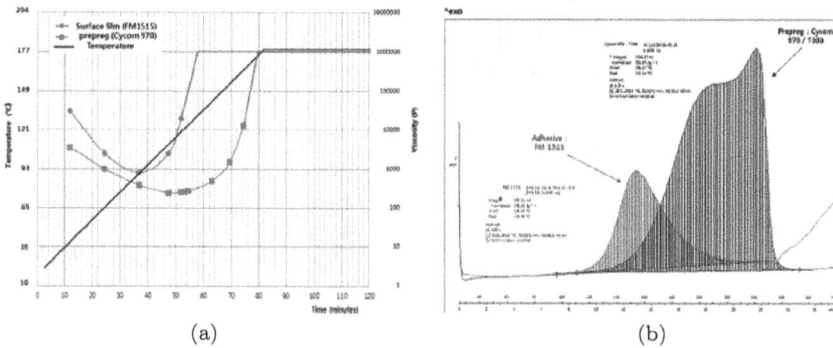

(a) (b)

Fig. 7.6 Material evaluation (a) viscosity profile (b) thermal analysis result (with permission).

With these results, material evaluations were carried out to check how to react between prepreg and surface film in cure stage. Viscosity and thermal analysis were done by DSC for surface film and carbon fiber prepreg, respectively. Figure 7.6 shows from viscosity analysis that when surface file on the outer layer reached gelation stage, carbon prepreg was still in the liquid stage. These different stages in the range of certain temperatures might entrap air under the surface film. For further investigation, as shown in Fig. 7.6, thermal analysis was performed. These investigations also showed that the surface film chemically reacted faster than carbon prepreg. This result also means that faster chemical reaction of the surface film does not give enough time to evacuate entrapped air through debulk process.

Finally, we can reach a certain conclusion that surface film on the outer layer, which is very often used for surface improvement in aerospace division, acts negatively like a barrier to prevent void evacuation and keeps voids locked in the laminate during the curing process. The influence of a surface film on void behavior in the laminate is evaluated and verified through complex-shaped sandwich structure test article fabrication.

7.3 VBO Process

VBO process is a method to fabricate parts with only vacuum pressure unlike the conventional autoclave method, and the material which is used for the VBO process has different characteristics from the prepreg for autoclave (A/C) process. As shown in Fig. 7.7, the feature of the partial impregnated material has the evacuation channel, which allows to remove air or moisture smoothly even in the vacuum stage.

The conventional autoclave process can surpass internal void easily because of the extra pressure (5–6 bar). However, VBO process removes the internal voids through the evacuation channel because of the vacuum pressure. To maintain the air path under the vacuum condition, as shown in Fig. 7.8, breather dam shall be applied to vacuum bagging.

7.3.1 *Case Study of Effect of Preprocessing for Efficient VBO Process*[10]

VBO process method has gained the most interest by offering energy efficiency using only an atmospheric pressure differential (1 atm) for part consolidation.[11] However, in the absence of high consolidation pressure, VBO prepreg must undergo compaction for longer

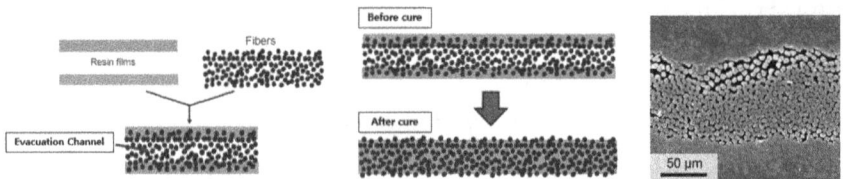

Fig. 7.7 The characteristic of the VBO material (CYCOM 5320-1).[9]

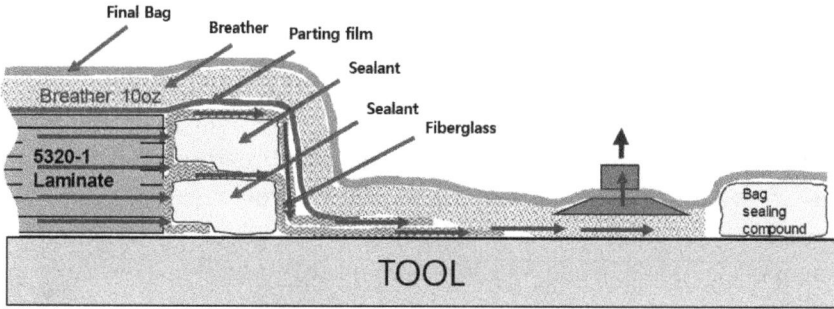

Fig. 7.8 Breather-dam application in VBO process.[9]

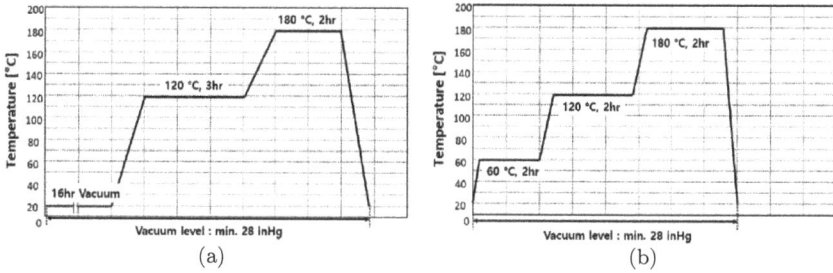

(a)   (b)

Fig. 7.9 The used cure cycles of (a) Referenced Method and (b) Modified Method.

durations during the curing process and requires the use of more elaborate processing schemes to conform to complex geometries. However, the recommended cycle by the material company involves a 16 h or more of vacuum hold in room temperature to evacuate gases trapped within the laminate during lay-up, followed by an initial cure at 121°C and post-cure at 177°C.[9] The objective of this research is modification of VBO prepreg cure cycle by reducing the overall cure cycle time while maintaining high part quality.

The aircraft bulkhead was selected as a sample demonstrator part which was fabricated using each cure cycle. The used cure cycles for comparison are of two types as shown in Fig. 7.9. The reference method, developed mainly for A/C processing, consists of a vacuum hold at room temperature for 16 h, followed by a ramp of approximately 1.5°C/min. Modified method, adding 2 h dwell at

$60°$C, is a recent method where the manufacturer has released a new cure cycle to reduce the total processing time by replacing the extended RT vacuum hold.

In order to observe the effect of preprocessing, the bulkhead parts were fabricated with referenced and modified method, respectively. As shown in Fig. 7.10, the main features of the bulkhead had complex configurations and consisted of less than 10 CFRP plies using CYCOM5320-1 T650 PW (plain weave) from Solvay with drop-off and honeycomb core areas. And it was 0.8 m $\times$ 0.3 m in size.

The internal defects were inspected from the viewpoints of wrinkle, overall void and micro void using section cut under microscope and digestion method according to EN2564[12] after cured parts with reference and modified method, respectively. Table 7.1 shows the result of porosity analysis for three test specimens extracted from the same locations of each part.

From the results, it turns out that the overall void content was significantly higher, approximately three times, in the reference method than in the modified method. Around 21 different specimens were selected from each demonstrator part (Figs. 7.11–7.12) and

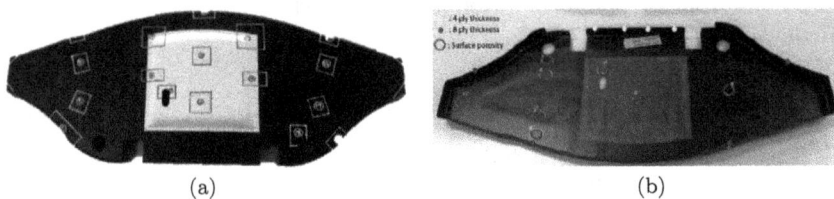

(a)   (b)

Fig. 7.10   Configuration of bulkhead structure of UAV. (a) Bag Side and (b) Tool Side.

Table 7.1   Void contents in bulkhead demonstrator part

	Reference (%)	Modified (%)
No. 1	2.78	1.54
No. 2	3.87	1.19
No. 3	3.52	0.76
Avg.	3.39	1.16

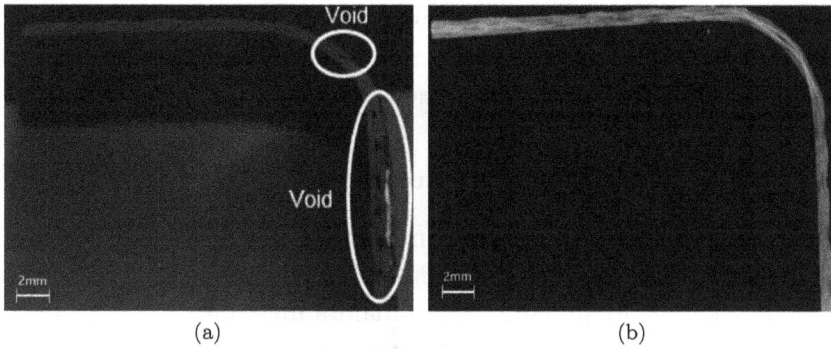

Fig. 7.11 Sectional view of cornered edge from bulkhead parts manufactured with (a) reference and (b) modified method.

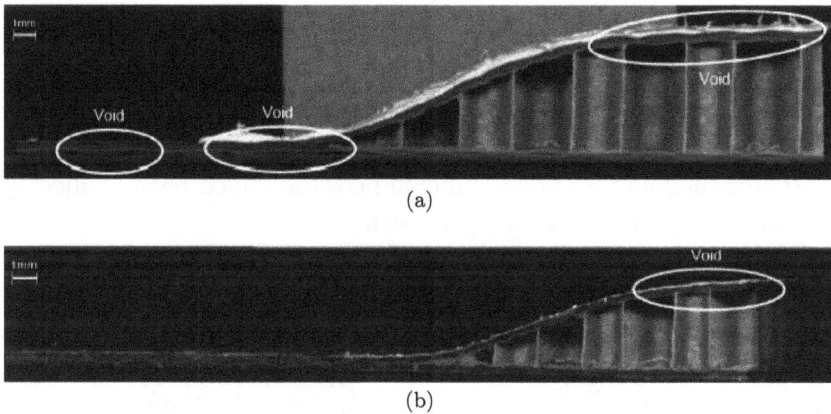

Fig. 7.12 Sectional view of sandwich structure facesheet from bulkhead parts manufactured with (a) reference and (b) modified method.

microscopy inspection was performed. Internal wrinkle was not observed, but the internal voids existed in the only corner area fabricated by the reference method. Additionally, the viscosity model was performed and it showed that the resin viscosity decreases by two orders of magnitude when the temperature is raised from RT to 60°C. Therefore, introducing a preprocessing step at an elevated temperature of 60°C instead of a vacuum hold at room temperature for 16 h is more effective for volatiles to escape under moderate flow level condition. The 2 h preprocessing step at 60°C, which is well below the onset of degree of cure rise, allowed voids to escape prior to breakdown of evacuation channels.

7.4 RTM (Resin Transfer Molding)

RTM is a process of injecting a mixed resin into a closed mold of a desired shape containing fiber reinforcement preform and then applying heat and pressure while maintaining inside of the mold under vacuum.[13] After injecting the resin, exothermic reaction occurs inside the mold and then the viscosity of the resin increases as time goes by. In the end, the resin is solidified. As shown in Fig. 7.13, the process of RTM can be divided into four steps as follows[14]:

(1) Place fiber reinforced preform which is similar to the desired shape of part inside the mold.
(2) Close the mold and seal tightly. The preform shall be carefully designed not to go beyond the sealed area of the mold.
(3) Once the mold is closed and clamped, the mixed resin is injected into the closed mold under pressure.

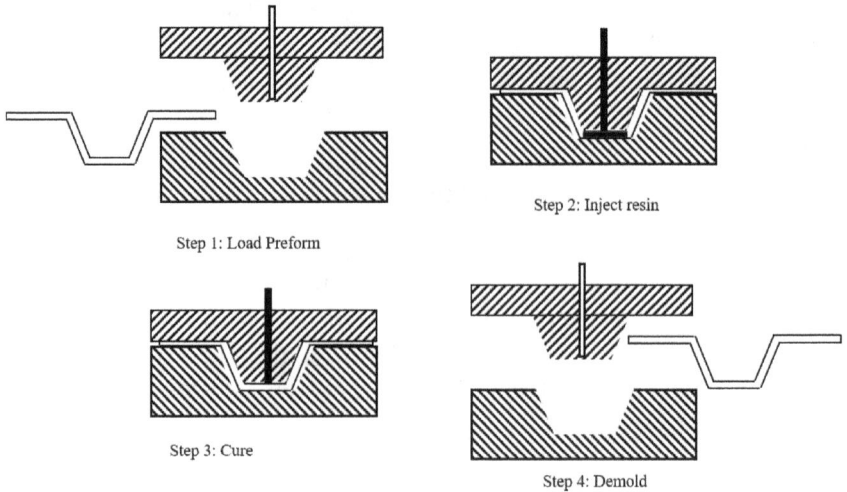

Step 1: Load Preform

Step 2: Inject resin

Step 3: Cure

Step 4: Demold

Fig. 7.13 Schematic diagram of the RTM process.

(4) Cure the part at an elevated temperature in the mold under the pressure and demold the cured part. If necessary, perform post-curing under vacuum to remove thermal stress and ensure mechanical properties.

RTM process has an advantage of using a low-cost tooling system compared to other processes such as injection molding or compression molding, because the pressure applied during the RTM process is lower than ones required in injection or compression molding. Generally, a mold used in RTM process is made of aluminum or steel. The mold is divided into two parts, which are one or multiple inlets where resin is injected and one or multiple outlets where air or excess resin is emitted. A single port is used for a small part but, a long and large part uses multiple ports for uniform resin distribution and quick process. In general, the inlet port on the lowest area of the mold and the outlet on the highest of the mold allow resin to flow against gravity and minimize the void. The thickness of the mold shall be strong enough to support all applied pressure during the process. Tool handling and thermal effect shall be fully considered during the design of the mold. The difference in thermal expansion between the mold and the fiber reinforced preform effects the dimension of the final part. So, the difference of coefficients of thermal expansion (CTE) shall be taken into account during the design of the mold. Resin flow and fiber wet-out are very important in the RTM process. The resin flow in the RTM mold depends on various parameters which are the injection pressure, the vacuum pressure inside mold, the temperature of the resin, viscosity, permeability of the preform, etc. The permeability is also affected by fiber type, fiber form, fiber volume fraction, etc.

7.5 Thermoplastic Composite Technology[15]

7.5.1 *Thermoplastic Composite*

Thermosetting resin is mainly used with the autoclave process where the prepreg is placed within the mold using hand lay-up or automated lay-up and then applying pressure and heat after placing it under a vacuum in the autoclave as per the curing process. The disadvantages of the autoclave process are its long process time and the difficulty of reworking after curing.[16] Unlike thermoset resin, thermoplastic resin can be produced with the desired part through the process that consists of (1) preheating a flat blank up to process temperature, (2) transferring to a press containing dies of desired shapes and (3) pressing until it cools below Tg.[17] It is similar to thermoset compression molding process using the conventional hot plate press. However, it is possible to thermoform using the thermoplastic blank in a short time. Therefore, it is low cost and can be easily mass-produced. The advantages of the thermoplastic resin in terms of functions and manufacturing are shown in Table 7.2. The typical high-performance thermoplastic resins applicable to aerospace division are polyetherketoneketone (PEKK), polyetheretherketone (PEEK), polyphenylenesulfide (PPS) and PEI (polyetherimide).[18]

These semi-crystalline resins which can be melted by heat have high mechanical properties, chemical resistance and are flame retardant. Especially, there are lesser deteriorations in mechanical properties (strength, stiffness, elongation, creep resistance and fatigue resistance) over a wide temperature range and they are highly effective in enhancing properties of continuous reinforced carbon or glass fiber.[19] In addition, they have high damage resistance, high operating temperature and low moisture absorption. They can be

Table 7.2   Thermoplastic's characteristics

Performance	Manufacturing
• Tougher, good fatigue performance	• Infinite storage life
• Damage tolerance	• Eliminate bagging materials and labor
• High-temperature performance	• Fusion bonding eliminates fasteners and adhesives
• Very low flammability, smoke, toxicity	• Short process cycle (1–30 min)
• Excellent chemical resistance	• Can be reformed
• Insensitive to moisture	• Recyclable, No VOCs (Volatile Organic Compounds)

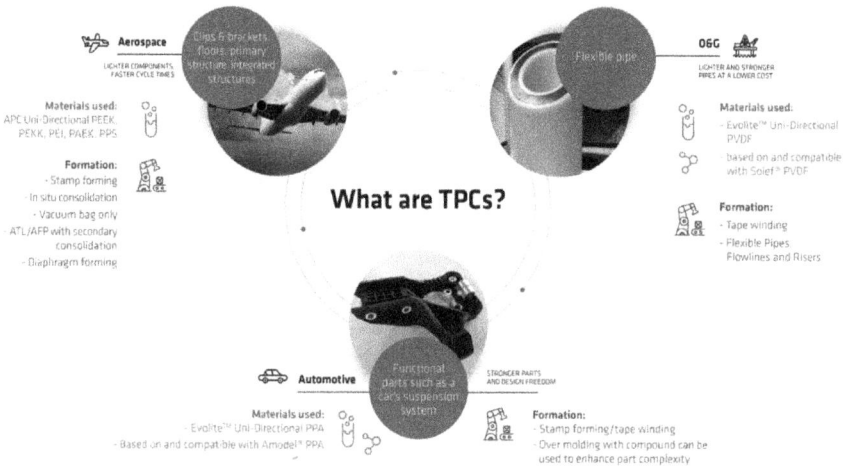

Fig. 7.14 Application of thermoplastic composites.[20]

applied to aerospace composite parts. But, high process temperature (350–400°C) and high viscosity make them harder to apply in various ways. The utilization of thermoplastic resin is shown in Fig. 7.14.

Currently thermoplastic resins for aircraft are mainly researched and developed on channels for interior reinforcement and airframe parts. Interior reinforcement is in the stage of mass production application and the manufacturing process for airframe parts such as landing gear doors, torsion box, etc., is under development at the level of TRL 4–6. Especially, interior parts must meet the requirement of Flame, Smoke, Toxicity (FST) to lower the risk of fire. So, it is highly suitable to apply high heat-resistant resins such as PPS and PEI for such parts. The high-performance resins such as PEKK, PEEK and PAEK are being studied for application in main aircraft structure.[21, 22]

7.5.2 *Thermoforming Process*

Thermoforming process allows low-cost and mass production for small parts with uniform cross-section such as ribs, clips and brackets in airframes. It is required to develop the technology for minimizing

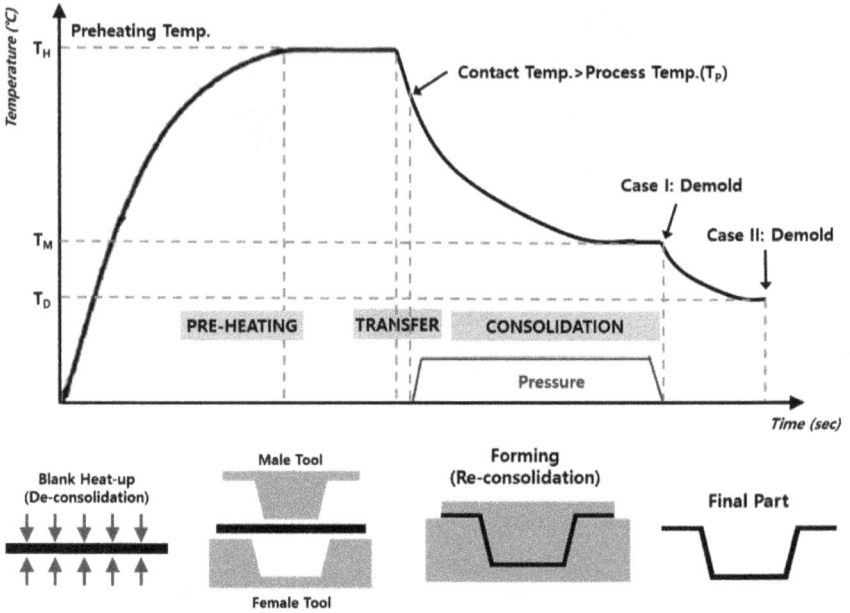

Fig. 7.15 Forming cycle for thermoplastic composite.

fiber distortion, dimensional instability and residual stress. Generally, as shown in Fig. 7.15, forming cycle for thermoplastic consists of (1) preheating, (2) transfer, (3) forming and (4) consolidation by cooling. It is possible to apply automation system to the entire process, which leads to the minimization of production unit price by mass production. By identifying process variables per process, a control system can be established to automate the entire process.

7.5.2.1 *Preheating*

This process keeps preconsolidated blanks at process temperature. It is required to heat above the melting temperature of the material and the temperature of the entire material shall be uniformly distributed in the thickness and length direction. If the forming is carried out in the state where the material does not reach the process temperature, internal defects such as matrix cracking, fiber buckling, fiber bridging and wrinkle might occur due to uneven resin flow inside the ply.

Table 7.3 Melting and process temperature for thermoplastic

Polymer	Crist	Glass transition temp. (°C)	Melting temp. (°C)	Process temp. (°C)	Maximum preheating temp. (°C)
PEEK	SC	140–145	334–343	340–390	425 ± 10
PPS	SC	85–95	275–290	315–345	330 ± 10
PEI	A	215–220	—	340–360	360 ± 10

On the contrary, if the heat temperature is too high, the desired quality cannot be acquired because of material degradation or resin overflow. Typical melting and process temperature for thermoplastic is shown in Table 7.3.

7.5.2.2 *Transfer*

The heated material shall be transferred to a press within approximately 3–5 sec which is not out of the range of the process temperature. Therefore, the preheating temperature shall be set based on the rate of heat loss per moving distance acquired through the experiment. The defects which may occur because the material temperature is higher or lower than the process temperature shall be prevented by setting the maximum preheating temperature while designing the process.

7.5.2.3 *Forming*

Forming is a process in which a heated material is accurately placed between the upper and lower mold of the press whereby the final shape is formed by applying pressure to the heated material. The temperature of the material shall maintain higher temperature than the process temperature until just before the forming. This process is the most critical stage which affects the thickness of the final shape, the surface condition, etc., and process variables such as the temperature of the upper and lower mold, pressure and moving speed must be controlled. As shown in Fig. 7.16, if the stamping rate is slow, the temperature of the material becomes lower than the process

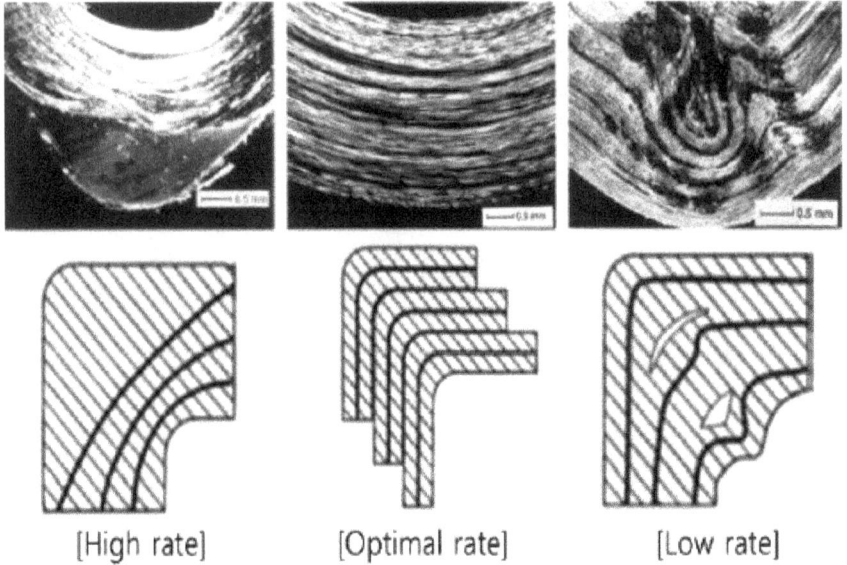

Fig. 7.16   Effect of stamping rate on the quality of bended part.[23]

temperature and then fiber buckling occurs because the slip between plies is not smooth.

On the contrary, if the stamping rate is too fast, the fiber gathers inside the bending area and the resins are collected on the outside by the speed difference between the fiber and resin flow that happens during the process. So the forming process could be finished imperfectly. The right moving speed should be applied according to the material and the final shape. Figure 7.17 shows the resin migration or fiber buckling that occurs when the stamping rate is slow or fast.

7.5.2.4 *Consolidation and Cooling*

The cooling rate is the most important parameter which affects the crystallinity of thermoplastic resin and the residual stress. Especially, in case of semi-crystalline thermoplastic resin, the higher the degree of the crystallinity, the better the mechanical properties and chemical resistance. However, the lower the degree of the crystallinity, the

In situ Consolidation

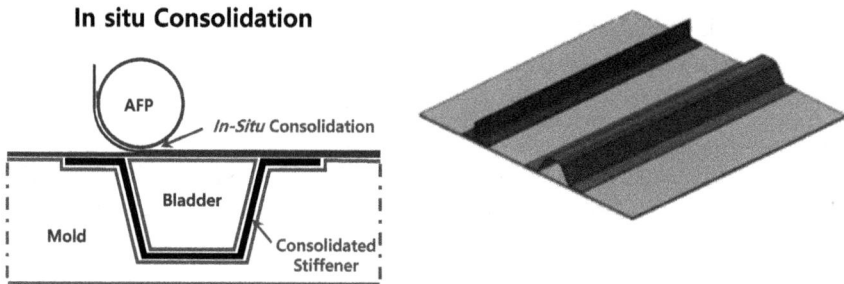

Fig. 7.17 Co-consolidation (*in situ* consolidation) by AFP.

lower the fracture toughness. In addition, the cooling rate is a crucial parameter which is determined by the temperature of thermoplastic blank and the mold. If the temperature of the mold is high, the cooling rate decreases and the degree of crystallinity increases.

7.5.3 Auto-Consolidation Technology

An auto-consolidation is a suitable technology which produces composite structures with thermoplastic resins at high speed. The pressure of the roller, lay-up speed, lay-up pattern and setting temperature of heat source shall be considered according to the characteristics of the raw materials. It has a mechanism similar to the automatic lay-up equipment. However, there is a difference in terms of the ability to manufacture products using *in situ* consolidation with a heat source such as a laser. In general, auto-consolidation technology has combined CAD technology with robot technology. In order to complete the fiber lamination structure given by CAD technology, the trajectory to be operated in software is converted into data and transferred to the robot, then the robot stacks the material along the trajectory. It can lower the unit price of manufacturing because of automation, repeatability, interoperability between design and manufacturing. The quality of the product is superior to the conventional hand lay-up method. This process can manufacture a product at high speed of up to 1 m/s using robot after installing Tow or Tape of various widths according to an equipment. This technology which can reduce the rate of scrap within 5% is suitable for manufacturing the skin part with large area among airframe structures.[24] Currently, this technology is mainly used for consolidated blanks, which are precollated ply packs before thermoforming process because of the limitation of technology for temperature control and *in situ* consolidation. As shown in Fig. 7.17, the research of co-consolidation process using *in situ* consolidation is in progress.[25]

7.6 Conclusions

Most of the primary structures in the aerospace division have been fabricated by the autoclave process because of its high mechanical property and dimensional stability. Due to the improvement in the mechanical and physical properties of raw materials used for the OOA process, the possibility of its application in primary structures is getting higher and higher. In fact, both the wings of the Bombardier C series (Now Airbus A220), a first in commercial airplane, have been fabricated by the vacuum infusion process of the OOA process with dry fiber composite. The OOA process introduced earlier has both pros and cons; therefore, it is essential to select an optimal process considering the shape and size of the products to be fabricated. Furthermore, cost competitiveness can be secured by transforming the optimized process into automation at a specific level.

References

[1] U. K. Vaidya and K. K. Chawla, *Mater. Rev.*, 53, 2008, 185.
[2] F. C. Campbell, *Manufacturing Technology for Aerospace Structural Materials*, Elsevier, 2006.
[3] T. Centea, L. K. Grunenfelder and S. R. Nutt, *Compos. Part A*, 70, 2015, 132.
[4] Abaris Training Resources, *Advanced Composite Structures*, Abaris, 1998.
[5] D.-C. Park and Y.-H. Kim, *Mod. Phys. Lett. B*, 33, 2019. 1940028 (5 pages).
[6] L. Liu, B. Zhang, Z. Wu and D. Wang, *J. Mater. Sci. Technol.*, 21(1), 2005, 87.
[7] L. Di Landro, A. Montalto, P. Bettini, S. Guerra, F. Montagnoli and M. Rigamonti, *Polym. Polym. Compos.*, 25(5), 2017, 371.
[8] X. Liu and F. Chen, *Eng. Trans.*, 64(1), 2016, 33.
[9] Cytec Engineered Materials. CYCOM 5320 information sheet.

[10] D. K. Hyun, D. Kim, J. W. Shin, B. E. Lee, D. H. Shin, J. H. and Kim, *J. Compos. Mater.*, Under Review.

[11] G. Bond, G. Hahn and J. Thomas, Non-Autoclave Manufacturing Technology: Demonstration of Applications, Defense Manufacturers Conference, Anaheim Convention Center Anaheim CA, 2011.

[12] EN 2564, *Carbon Fibre Laminates — Determination of the Fibre, Resin and Void Contents*, German Institute for Standardisation, 1998.

[13] D.-C. Park, T.-G. Kim, S.-H. Kim, D.-H. Shin, H.-W. Kim and J.-W. Han, *Compos. Res.*, 31, 2018, 304 (in Korean).

[14] F. C. Campbell, *Manufacturing Processes for Advanced Composites*, Elsevier, 2004.

[15] B. E. Lee, D. K. Hyun and D. H. Shin, *Transac. Mater. Proc.*, 27, 2018, 60 (in Korean).

[16] B. Vieille, W. Albouy, L. Chevalier and L. Taleb, *Compos. Part B*, 45, 2013, 821.

[17] A. R. Offringa, *Compos. Part A*, 27, 1996, 329.

[18] M. Biron, *Thermoplastics and Thermoplastic Composites*, Elsevier, 2013, 1.

[19] J. Diaz and L. Rubio, *J. Mater. Process Technol.*, 143, 2003, 342.

[20] https://www.solvay.com/en/chemical-categories/our-composite-materials-solutions/thermoplastic-composites.

[21] A. Deterts, A. Miaris and G. Soehner, Internation Conference & Exhibition on Thermoplastic Composites (ITHEC), 2012, p. 60.

[22] Dutch Thermoplastic Components (DTC), http://www.composites.nl/products.

[23] N. O. Cabreraab, C. T. Reynolds, B. Alcocka and T. Peijs, *Compos. Part A*, 39, 2008, 1455.

[24] Z. August, G. Ostrander, J. Michasiow and D. Hauber, *SAMPE J*, 50, 2014, 30.

[25] https://www.compositesworld.com/articles/thermoplastic-composite-wings-on-the-horizon.

Chapter 8

Application of Composite Materials for Shipbuilding and Marine Engineering

Sung-Won Yoon

Department of Advanced Materials,
Research Institute of Medium & Small Shipbuilding,
Busan, Republic of Korea
ysw8114@naver.com

International Maritime Organization (IMO) regulations on carbon dioxide and greenhouse gas emissions have come into force, requiring energy saving and eco-friendly material technology. Therefore, there is a need for a lightweight material that can maintain stability and durability in an extreme marine environment and which has excellent mechanical performance. As the IMO's environmental regulations have been strengthened to increase the energy efficiency of ships, the IMO has also begun to consider operational economics such as energy reduction through lightening the hull. Demand for lighter weight technology using composite materials is increasing. Examples would include lightweight large structures using composite materials, composite materials replacing metal design parts, and polymer composite materials applicable to marine environments. When the existing metal material is replaced with a composite material, the cargo transport volume increases due to weight reduction and the operational efficiency is improved, so a high economic effect can be expected.

The purpose of this study is to determine the correct estimation of laminate patterns for high-strength composite shafts applied to a ship. In this study, analysis is carried out on the application of a high-strength CFRP shaft. In the composite material, the mechanical properties vary depending on the laminated pattern of the reinforcing material. Therefore, in this study, the properties of the composite materials were calculated using a computer simulation program. The laminate patterns in the filament winding process are an important factor in determining the strength and life of the final structure. In this study, the structural safety was analyzed for the laminate patterns in four cases. The laminate patterns and the order of the layers were determined by considering the results of the finite element analysis. The shear stress equation of the hollow shaft for torsional loads showed that the thickness of the structure varied with the diameter ratio. An important design variable to be considered when designing composite material intermediate shafts is the natural frequency for resonance avoidance at critical rotational speed and torsional strength for axial load. In order to satisfy these, strength and modal analysis were performed. The stacking pattern and the stacking order were finally decided considering the results of the infinite element analysis (FEA).

8.1   Introduction

8.1.1   *Background for Application of Composite Materials in Shipbuilding and Marine Industries*

As part of responding to global environmental regulations, the shipbuilding and marine industries are increasingly interested in eco-friendly ship design and construction technology to improve ship operation efficiency of ships. Therefore, worldwide technology developments related to the use of eco-friendly fuel, in response to marine environment protection regulations, for energy savings and improvement of operational efficiency are actively carried out.

In addition, the development of design parts for eco-friendly ships using high-functional materials is underway, and the demands for development of hulls and equipment with characteristics such as high strength, light weight and corrosion resistance are continuously increasing. It is believed that the application fields of these high-functional materials are expanding to reduce the weight of transportation equipment such as automobiles and aircraft, and that the possibility of application to the next generation new industrial fields is high due to the advancement of the shipbuilding and marine industries. Table 8.1 shows the classification of composite material technology applied to shipbuilding and marine industries.[1–15]

In particular, fiber reinforced plastics which represent high-functional materials exhibit various anisotropies depending on the

Table 8.1 Classification of composite material technology applied to ship-building and marine industries

Classification	Details
Polymer	• Development of high-performance and low-cost resins for optimal production of polymer composite materials • Development of resins applicable to shipbuilding and marine environments (semi-non-combustible, moisture resistant and low-temperature environmental durability)
Reinforcement	• Reinforced fiber manufacturing technology to ensure economic feasibility • Lightweight core material and preform manufacturing technology • Reinforced fiber manufacturing technology applicable to shipbuilding and marine industries
Manufacturing process	• Composite material process technology applied to large structures • Advanced manufacturing process technology for high-speed production • Technology for bonding different materials (composite materials and metal materials)
Parts technology	• Part technology of lightweight composite material structures • Design and analysis technology of composite materials for structures • Internal and external parts and power train parts technology • Marine structure parts technology

Source: Project Planning Report for Industrialization of Marine Convergence Composite Materials.

stacking order and direction of the reinforcing materials, and have the advantages of securing the mechanical properties required for development of products. Further, composite materials have superior specific strength, specific rigidity and corrosion resistance compared to single metal materials, so their application is increasing in the shipbuilding and marine industries.[16–24]

Generally, ships with composite materials have excellent fuel efficiency and corrosion resistance, which reduces operating costs of ships. As a result, many shipbuilding companies are actively considering the application of composite materials for the low-weight design of hulls, decks and structures. Therefore, in this chapter, application examples of composite materials in the shipbuilding and marine industries are provided and the technology development trends are explored.

8.1.2 Application Examples of Parts and Systems of Composite Materials Applied to Shipbuilding and Marine Industries

Recently, the International Maritime Organization (IMO) has implemented carbon dioxide reduction and greenhouse gas emission regulations, so the shipbuilding and marine industries are required to develop eco-friendly ship design and construction technology. Therefore, the lightweight materials with excellent mechanical performance while maintaining stability and durability in extreme marine environments are needed. Figure 8.1 shows the application of composite materials in eco-friendly ships and offshore plants. Since 2013, IMO has enacted regulations to reduce carbon emissions by up to 30% and nitrogen oxide (NOX) emissions by 20% or more. In many international countries, it is mandatory to build eco-friendly ships subject to greenhouse gas emission regulations by accepting the international agreements to restrict the amount of carbon dioxide emitted from ships as domestic laws. Accordingly, the shipbuilding industry has started to develop eco-friendly ship technology, and the marine industry is in an urgent need to prepare countermeasures to reduce transportation costs.

In the interim, an economical manufacturing process should be possible for large structures, and composite materials are most likely to support this. The composite materials applied to shipbuilding and marine industries refer to high molecular materials with high structural performance, reinforced with organic and inorganic fillers, such as carbon fiber and glass fiber, having heat resistance and excellent refractory properties to satisfy the requirement performance. As a

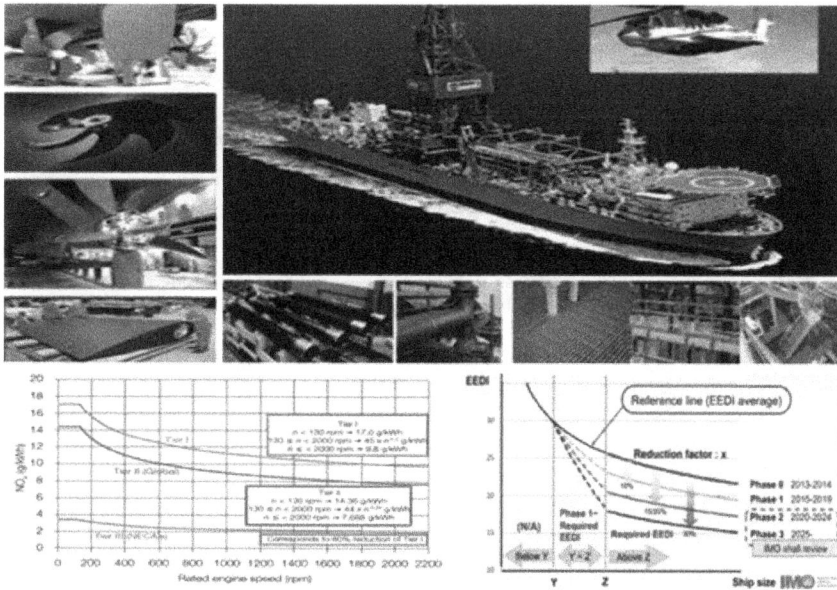

Fig. 8.1 Application of composite materials in eco-friendly ships and offshore plants.

Source: Project Planning Report for Industrialization of Marine Convergence Composite Materials

structural material applied to shipbuilding and marine industries, the polymer composite materials are already being researched and applied actively in Europe, where various shipbuilding and marine parts are developed and applied through multi-material parts development projects such as the European Network for Lightweight Application (E-LASS). The polymeric composite materials were mainly reported to have replaced structural parts such as deck, mast, wind wall, etc., of military vessels or high value-added ships. Various types of solutions are being suggested to apply these lightweight composite materials to large plants. Figure 8.2 shows the examples of the lightweight materials used for the global network construction in Europe's shipbuilding and marine industries (E-LASS).

In order to establish a system of cooperation with the shipbuilding industry to achieve competitiveness in lightweight parts and materials technology that is rapidly growing around the world, it is

Fig. 8.2 A case of the lightweight composites used for global network construction in Europe's shipbuilding and marine industries (E-LASS).

Source: Project Planning Report for Industrialization of Marine Convergence Composite Materials

necessary to proactively develop the core parts for shipbuilding and marine products based on high-functioning high-molecular composite materials. The manufacturing technology of composite materials in the field of high-functional materials is expanding its applications to make transportation equipment lighter. As a technology that is currently intensifying the competition in development due to the increase in demand, it is possible to increase the applicability of new industries for the next generation and contribute to the national economic development due to the advancement of existing industries.

8.1.3 *The Importance and Growth of High-Molecular Composite Materials for Shipbuilding and Marine Engineering*

As the IMO emphasized the energy efficiency of ships due to the strengthening of environmental and safety regulations, the industry began to consider the economic feasibility of operations such as energy reduction through the lightening of hulls. As the demand for lightweight technology for hulls increases, the development of the technology for large-sized composite materials, lightweight structures, replacement materials for metal design parts and polymer composites applicable to marine environments is rapidly rising. When

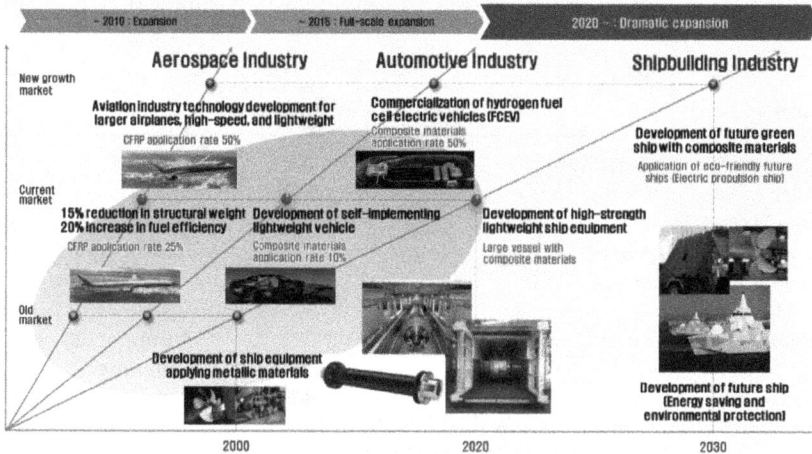

Fig. 8.3 Growth trends of composites in different industries.

the structural parts for shipbuilding and marine industries which are mostly made of metal materials are replaced with composite materials, it will increase the volume of cargo transport due to the lightening and improvement in operational efficiency, which will lead to high economic effects. Figure 8.3 shows the growth trends of composites in different industries.

8.1.4 Composite Material Propulsion Shaft Technology Applied to Shipbuilding and Marine Industries

The power shaft of the ship's propulsion system is composed of an intermediate shaft, a thrust shaft, a propeller shaft and a stern tube shaft. Most of them are made of metal materials such as forged steel. However, in the developed nations such as in the European countries, the intermediate shaft, propeller, stern tube bearings, etc., of the ship propulsion system have already been developed from fiber reinforced composite materials and applied to actual ships in recent years. Table 8.2 shows the economic effects of the superstructures of ships with composites. Compared to the existing metal material propulsion

Table 8.2   The economic effects of the superstructures of ships with composites.

Item	Superstructure		Remarks
	Steel	Composites	
Superstructure weight (tons)	950	440	
Hull Lightening weight (tons)	—	510	
Superstructure production cost (US million \$/year)	7	12	Material and labor costs included
Annual freight charges* (US million \$/year)	—	3.06	
Freight charges (US million \$)	—	76.5	Assuming 25-year lifetime
Breakeven point (year)	—	3.9	

Note: *Annual freight charges: As freight charge depends on the type of cargo to be transported, calculate with 500 \$/tons · months as average charge.

Source: Tommy Hertzberg, LASS, 2009.

shaft, shafts made of fiber reinforced composite material can achieve significant weight reduction, which can achieve remarkable effects such as improved operational efficiency and vibration reduction. In addition, the effects of propulsion shafts composed of fiber reinforced composite materials are as follows:

(1) Possible to reduce the weight by applying fiber reinforced composite material compared to the existing metal material propulsion shaft.
(2) Possible to reduce vibration noise by removing the torsional vibration avoidance area (resonance point) of the propulsion shaft system from the main engine operation area.
(3) Possible to reduce the torsion stiffness due to the high attenuation characteristic of the composite resin (base material).
(4) Possible to omit the shaft support bearings by the lightening of the propulsion shaft and the minimal necessary adjustment of bending stiffness.
(5) Possible to enhance the maintenance by excellent environmental and corrosion resistances of the fiber reinforced composite materials.

The metal propulsion shafts widely applied to general ships have been established in the classification rules of each country for design, manufacturing and evaluation technologies. It is considered that the technology and reliability have been accumulated for a long time. However, the technology for design and evaluation to apply to the propulsion shaft of a fiber reinforced composite material has no special regulations in other classifications except the DNV-GL classification. DNV-GL classification so far has only established the rules for the twin-axle of small and medium-sized ships, and it is necessary to establish the rules and regulations of technology for the design and evaluation of large-sized ships and main shafts. Figure 8.4 shows the basic structure of the propulsion shaft.

The propulsion shaft installed on the ship can be manufactured using composite material molding technology such as filament winding, which is one of the molding methods of fiber reinforced composite pipes, pressure vessels, etc., in order to improve performance in terms of light weight, low vibration and convenience in maintenance.

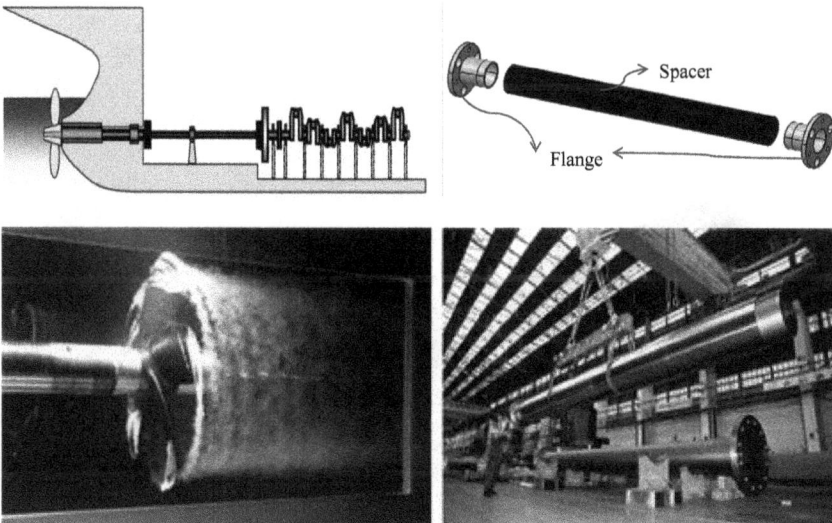

Fig. 8.4 The basic structure of the propulsion shaft.

8.1.5 Research Purpose and Contents

In this chapter, we study the feasibility of optimal structure design
for a fiber reinforced composite material propulsion shaft that has
advantages in operational efficiency, such as low vibration, compared
to the propulsion shaft of metal materials applied to the existing
large ships. Figure 8.5 shows the shape of the propulsion shaft.

Case 1: $[\pm15, \pm30, \pm30, \pm45, \pm45, \pm45, \pm45, \pm30, \pm30, \pm15]15$

Case 2: $[\pm15, \pm30, \pm45, \pm45, \pm75, \pm75, \pm45, \pm45, \pm30, \pm15]15$

Case 3: $[\pm30, \pm30, \pm45, \pm45, \pm75, \pm75, \pm45, \pm45, \pm30, \pm15]15$

Case 4: $[\pm30, \pm30, \pm45, \pm45, \pm60, \pm60, \pm45, \pm45, \pm30, \pm15]15$

Currently, the metal material propulsion shafts that are widely
applied to general ships are defined in the classification rules of
each country for design, manufacturing and evaluation technologies,
but there is no technology for design and evaluation to apply to
propulsion shafts of fiber reinforced composite materials.

In this study, for the optimal design of the propulsion shaft
from the fiber reinforced composite materials for large ships, the
optimal dimension and stacking angle for the propulsion shaft tube
are derived by using the critical angular velocity and shear stress
formulas commonly applied to propulsion shaft design.

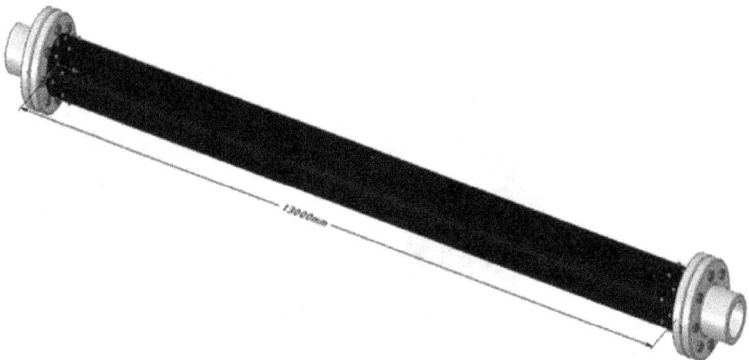

Fig. 8.5 A shape of the propulsion shaft.

In order to examine the validity of the design conditions of the propulsion shaft tube derived from this study, computerized analysis was performed based on the theory proposed by Tsai-Wu to verify the structural breakdown pattern, and based on these results, the propulsion shaft for fiber reinforced composite material for large ships was manufactured in a reduced scale to evaluate various mechanical properties and to evaluate the effectiveness of the design.

8.2 Design of the Composites Intermediate Shaft

The design of the existing metal propulsion shaft for ships is generally planned and manufactured according to the rules prescribed by the classification society of each country, and then approved and mounted on the ship. However, it is impossible to obtain approval according to the existing classification rules because the use of the composite material propulsion shaft targeted in this study is very rare. For this reason, the design of the composite material propulsion shaft must be designed and manufactured under stricter procedures than the existing procedures to ensure stability.

In this chapter, we will determine the dimension of the composite material tube and derive the stacking angle for optimal design of the propulsion shaft for the application of the composite material. To do so, the shear stress distribution of the shaft according to the ratio of the diameter to the hollow shaft was examined to derive the optimal diameter ratio, and the shear strength was calculated according to the diameter ratio. After that, an arbitrary stacking angle was selected within the stacking angle range of $-75° \leq \theta \leq 75°$ to check the shear strength for the laminated composite specimen through MCQ simulation results.

8.2.1 *Determination for the Dimension of the Hollow Shaft Tube Made of the Composite Material According to the Diameter Ratio*

In order to secure the reliability of the composite tube for the propulsion shaft made of a hollow shaft, a process for checking the validity of the tube by examining the diameter of the tube calculated using the dimension determination theory must be preceded.

Therefore, in this study, the design validity of the composite propulsion shaft was examined through the critical angular velocity and shear stress equations applied to the well-known composite shaft. And then, we maximized the performance of the shaft system through design optimization.

8.2.2 *Dimension Decision Theory*

For the dimension design of the propulsion shaft of the composite material, the diameter of the hollow shaft with a length of 13,000 mm, which is the target propulsion shaft of this study, was calculated through the equation of the critical angular velocity generally known. The equation of the critical angular velocity is specified below and a graph showing the relationship between the shaft diameter and the length of the shaft calculated by the equation is shown in Fig. 8.6. The critical angular velocity formula is a commonly used formula for small and medium-sized composite propulsion shafts. When applied on the basis of a large propulsion shaft of 13 m, it requires a large

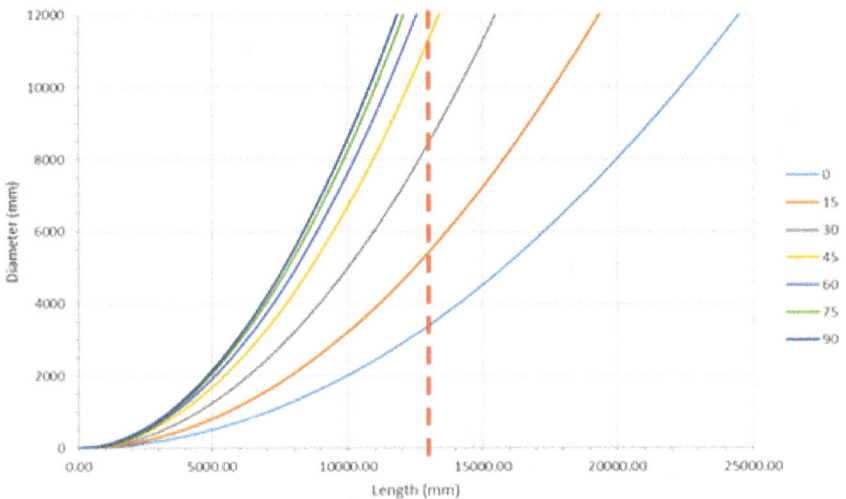

Fig. 8.6 Relationship between the length and diameter of shaft according to winding angle.

shape with a diameter of at least 3 m. Therefore, we confirmed that it is unsuitable as a formula for the design of propulsion shafts for large vessels.

$$L_{\text{crit}} = \sqrt{\frac{\pi}{4\sqrt{2}} \frac{d_m}{V_{\text{crit}}} \sqrt{\frac{E_x}{\rho}}} \qquad (8.1)$$

$$V_{\text{crit}} = \frac{\pi}{4\sqrt{2}} \frac{d_m}{L^2} \sqrt{\frac{E_x}{\rho}} \qquad (8.2)$$

$$E_x = \left[\frac{1}{E_1}\text{Cos}^4\theta + \left(\frac{1}{G_{12}} - \frac{2V_{12}}{E_1} \right) \sin^2\theta\cos^2\theta + \frac{1}{E_2}\sin^4\theta \right]^{-1} \qquad (8.3)$$

here, each sign refers to as follows:

L_{crit}: Critical length of tube
d_m: Average diameter of tube
V_{crit}: Critical angular velocity
L: Length of tube (distance between both end supports)
ρ: Density of composite material used for tube
E_x: Young's modulus of tube along axis direction
E_1, E_2: Young's modulus of fiber direction and perpendicular direction of the fiber
G_{12}: Shear stiffness

As mentioned previously, when checking the graph of the relationship between the shaft length and the diameter according to the stacking angle, the shaft diameter increases steeply as the length of the shaft increases, even if the slope of the diagram is the slowest $0°$ stacking angle criterion. Therefore, it was confirmed that it was impossible to apply to the design of the target large shaft, which was the target in this study. In addition, even if this formula is used, it is difficult to derive the optimum thickness of the composite tube with respect to the hollow shaft, so it is estimated that it is difficult to be utilized in this study. Accordingly, in order to obtain the optimum thickness, the design of the composite material propulsion shaft was performed by applying the shear stress equation of the shaft system as an alternative to the critical angular velocity equation.

8.2.3 *Derivation of Optimum Diameter Ratio*

Due to the results that the critical angular velocity equation commonly used in the design of composite materials hollow shaft is not suitable for this study, the shear stress equation for the shaft system was reviewed to derive the dimensions. The shear stress equations for the shaft system are shown in Equations (8.4) and (8.5). The most important factor in designing the composite material propulsion shaft as a hollow shaft is to derive the optimum thickness of the composite material tube. In the manufacturing process of the composite material propulsion shaft, when the thickness of the layer is too much, the base material impregnated with the fiber does not harden, which may lead to process failure. Therefore, it is important to select the thinnest from the design process. However, if the thickness is too thin, it may cause damage to the structure due to torsion and buckling.

$$\tau_{xu} = \frac{16T}{\pi D_0^3 (1 - R^4)} \tag{8.4}$$

$$(\tau_{xu})_M = \bar{\bar{\tau}}_{xu} \tag{8.5}$$

$$T(\text{kNm}) = 9.5488 \times P(\text{kW})/S(\text{RPM}) \tag{8.6}$$

here, each sign refers to as follows:

τ_{xu}: Shear stress of tube
T: Torque of tube
D_0: Outer diameter of tube
R: Ratio of outer diameter to inner diameter of tube.

Using the shear stress equation of the hollow shaft, the change in shear stress according to the diameter ratio was derived as shown in Figs. 8.7 and 8.8. The factors used in the design of the shaft are listed in Table 8.3. In this study, the final goal was to design a propulsion shaft for a maximum torsional load of 1,800 kN·m. In order to secure the reliability of such a shaft, a safe propulsion shaft must be designed even at a torsional load of 4,500 kN · m calculated using Equation (8.6). Therefore, in this chapter, considering the two loads and reflecting them in the design of the shaft, the thickness of the shaft was selected by setting the outer diameter of the shaft in the

Variation of shear stress with the ratio of dimension

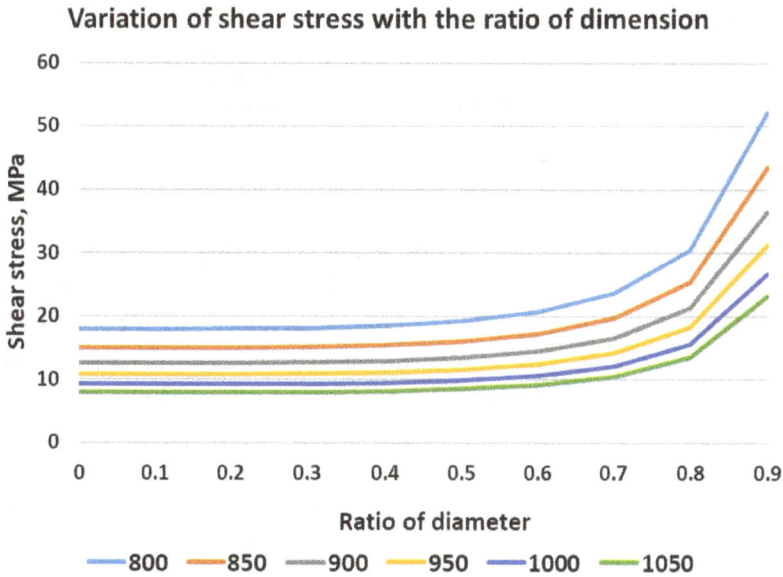

Fig. 8.7 Variation of shear stress with the ratio of dimension (1,800 kN · m).

Variation of shear stress with the ratio of dimension

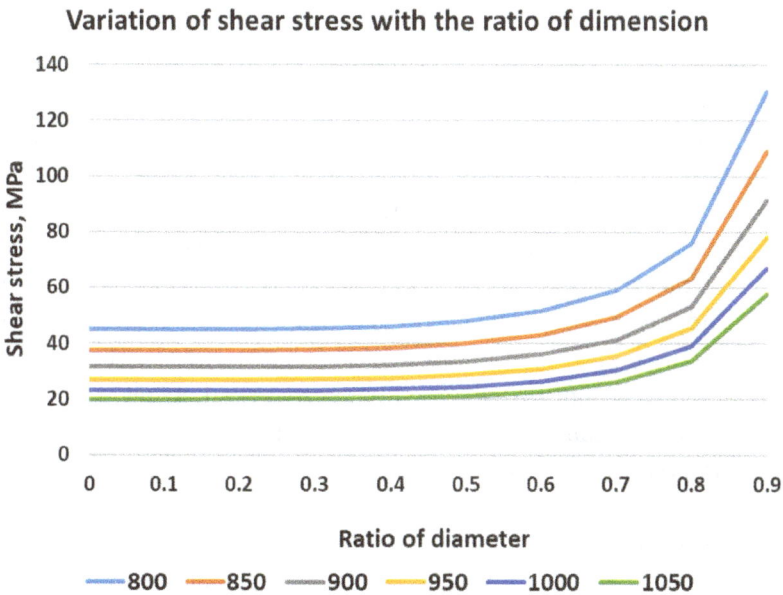

Fig. 8.8 Variation of shear stress with the ratio of dimension (4,500 kN · m).

Table 8.3 Design condition of composite material propulsion shaft

Speed (RPM)	70.2	Power (kW)	33,000
Torque (kN · m)	1,800/4,500	R (Diameter Ratio)	0∼0.9
D$_0$ (Outer diameter)		800, 850, 900, 950, 1000, 1050	

range of 800–1,050 mm, and considering the ratio of the diameters to the total six outer diameters and the shear load according to them.

Referring to the graph of the change in shear stress compared to the diameter ratio according to the outer diameter of the propulsion shaft, it showed a constant shear stress result without significant change up to the diameter ratio of 0.4, but gradually increased from the diameter ratio of 0.5, and it was confirmed that the stress increased rapidly as it passed 0.8 of the diameter ratio. It is generally known that structural stability can be secured when the diameter ratio of the hollow shaft is 0.4. However, the propulsion shaft targeted in this study is applied to large ships, so it is a condition that the actual thickness of the tube is at least 200 mm, which makes it impossible to actually manufacture such a large shaft.

Accordingly, in order to examine the conditions for actual manufacturing of the shaft by deriving the tube thickness compared to the diameter ratio according to the diameter of the shaft, the shear strength by thickness was summarized in Figs. 8.9 and 8.10. As a result of deriving the thickness of the tube compared to the diameter ratio, the tube thickness was confirmed to be 100 mm or less at the diameter ratios of 0.8 and 0.9, and selected as a suitable diameter ratio. The reason why the thickness of the tube by diameter ratio is less than 100 mm is that if the thickness is too much, the designed performance may not be exhibited because a rigid structure does not form on the base material impregnated with the reinforcing material, so a rigid structure is not formed.

To select an optimized diameter ratio for diameter ratios of 0.8 and 0.9, the thickness and required shear strength were summarized by classifying the target torque and the maximum torque for the diameter of the propulsion shaft of 900 mm targeted in this study. It was estimated that the required shear strength was 34 MPa when the

R	800 I.D	800 T	850 I.D	850 T	900 I.D	900 T	950 I.D	950 T	1000 I.D	1000 T	1050 I.D	1050 T
0.1	80	360	85	383	90	405	95	428	100	450	105	473
0.2	160	320	170	340	180	360	190	380	200	400	210	420
Original R Range												
0.3	240	280	255	298	270	315	285	333	300	350	315	368
0.4	320	240	340	255	360	270	380	285	400	300	420	315
0.5	400	200	425	213	450	225	475	238	500	250	525	263
0.6	480	160	510	170	540	180	570	190	600	200	630	210
Optimal R Range												
0.7	560	120	595	128	630	135	665	143	700	150	735	158
0.8	640	80	680	85	720	90	760	95	800	100	840	105
0.9	720	40	765	43	810	45	855	48	900	50	945	53

Fig. 8.9 Ratio of the outer diameter and inner diameter.

R	900 (1,800 kNm) I.D	t	τ_{xy}	900 (4,500 kNm) I.D	t	τ_{xy}
0.1	90	405	13	90	405	31
0.2	180	360	13	180	360	32
Original R Range			13	270	315	32
0.4	360	270	13	360	270	32
0.5	450	225	13	450	225	34
0.6	540	180	14	540	180	36
			17	630	135	41
Optimal R Range						
0.8	720	90	21	720	90	53
0.09	004	40	34	004	40	84
0.9	810	45	37	810	45	91

Fig. 8.10 Shear strength by thickness.

torque was 1,800 kN · m, and the required shear strength was 64 MPa when the torque was 4,500 kN · m.

8.2.4　*Determine the Optimal Stacking Angle*

This chapter intends to establish the design optimization of the composite material propulsion shaft by conducting a review of the optimal diameter ratio and tube thickness derived from the previous research results. Since the stacked structures such as the fiber reinforced composite materials exhibit various properties according to the stacking angle of the reinforced materials, the stress character-istics of the structure to be developed should be carefully examined to select the appropriate angle to maximize the characteristics of the design flexibility, which is the greatest advantage of the composite materials. In order to check to what extent the shear strength, which is the minimum required property according to the thickness of the tube compared to the diameter ratio, is satisfied at an arbitrary selected stacking angle, the simulation was performed using MCQ-Composites, a property analysis program for composite materials. Based on the simulation results, the mechanical properties of the composite stacking structure were confirmed and reflected in the design. The reason for using the simulation without actual specimen test to check the properties of the composite material is that the carbon fiber material used in this study is formed at an expensive price and the tube thickness applied to the large propulsion shaft is at least 48 mm, so it is impossible to make actual specimens. Also, it was considered in a reasonable way to calculate the properties by directions using simulation, in order to derive the properties for application to the structural analysis.

8.2.5　*Determination of Stacking Angle of Composites Propulsion Shaft Tube*

In composite materials structure, the determination of stacking angles is the most important factor for designing structures, and even a structure of the same thickness may vary depending on the stacking angle. In order to determine the appropriate stacking angle of the composite material propulsion shaft, the mobility of the propulsion shaft should be reviewed, and then the direction in which the stress is transmitted should be analyzed. So that the stability of the structure can be ensured. Therefore, in this chapter, the change in elastic modulus compared to the stacking angle is plotted as a graph as shown in Figs. 8.11 and 8.12 to analyze the elastic modulus according to the stacking angle of the reinforcing material and use it to determine the optimal stacking angle. The change in elastic modulus compared to the stacking angle was confirmed by the following equations:

$$\frac{1}{E_{\text{xlamina}}} = \frac{1}{E_{11}}C^4 + \left[\frac{1}{G_{12}} - \frac{2v_{12}}{E_{11}}\right]S^2C^2 + \frac{1}{E_{22}}S^4 \tag{8.7}$$

$$\frac{1}{E_{\text{ulamina}}} = \frac{1}{E_{11}}S^4 + \left[\frac{1}{G_{12}} - \frac{2v_{12}}{E_{11}}\right]S^2C^2 + \frac{1}{E_{22}}C^4 \tag{8.8}$$

$$\frac{1}{G_{\text{xulamina}}} = 2\left[\frac{2}{E_{11}} + \frac{2}{E_{22}} + \frac{2v_{12}}{E_{11}} - \frac{1}{G_{12}}\right]S^2C^2 + \frac{1}{G_{12}}[C^4 + S^4] \tag{8.9}$$

As a result of the change in elastic modulus compared to the stacking angle in the graph, it was estimated that it was most ideal

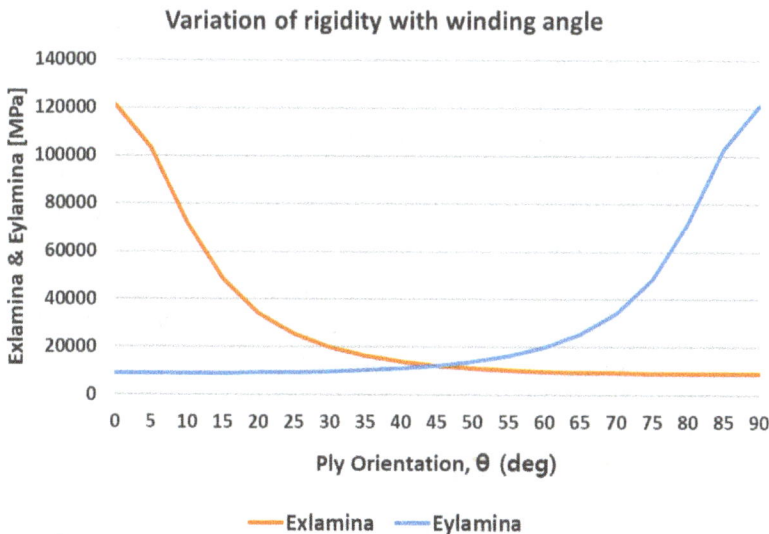

Fig. 8.11 Variation of the E_{xlamina}, E_{ylamina}.

Gxylamina Vs. Ply Orientation

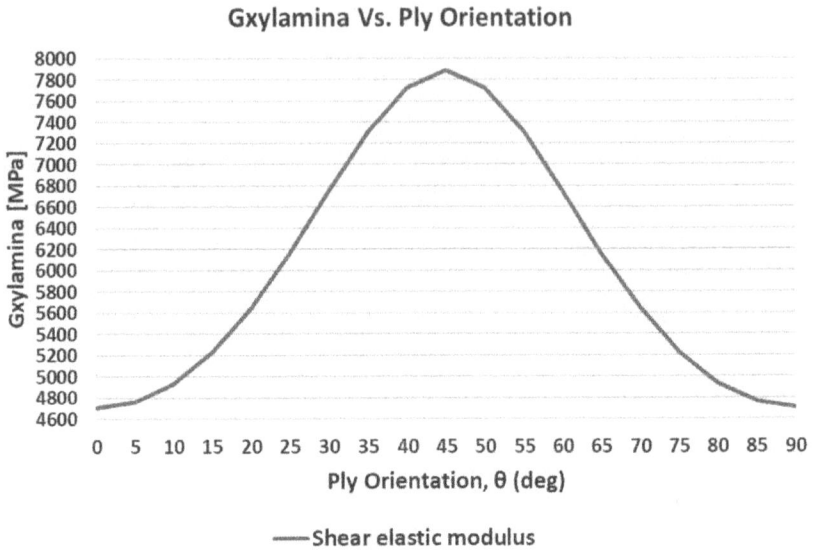

Fig. 8.12   Variation of the $G_{xylamina}$.

to be designed in the direction of 45° to match the motion performance of the structure subjected to torsional load. However, this
result is considered only in the theoretical formula. Therefore, it is
estimated that the method for stacking with other angles would be
considered, referring to the length and diameter of the shaft. For
more accurate determination of the stacking angle, the relationship
between stress and strain was reviewed to select the range of the
stacking angle, and then the stacking pattern was derived. The graph
of the relationship between the stress and the strain according to the
stacking angle is shown in Figs. 8.13–8.16. According to the results of
the review, the shear stress and the strain were the most stable in the
range of the stacking angles from 15° to 75°. Considering the length
and diameter of the composite material propulsion shaft based on the
aforementioned results, the stacking angle was selected in the range
of 15°–75° to construct the random stacking patterns corresponding
to four cases totally, which are as follows:

Case 1: $[\pm 15, \pm 30, \pm 30, \pm 45, \pm 45, \pm 45, \pm 45, \pm 30, \pm 30, \pm 15]$ ply number
Case 2: $[\pm 15, \pm 30, \pm 45, \pm 45, \pm 75, \pm 75, \pm 45, \pm 45, \pm 30, \pm 15]$ ply number

Case 3: $[\pm30, \pm30, \pm45, \pm45, \pm75, \pm75, \pm45, \pm45, \pm30, \pm30]$ ply number

Case 4: $[\pm30, \pm30, \pm45, \pm45, \pm60, \pm60, \pm45, \pm45, \pm30, \pm30]$ ply number

8.2.5.1 *Relationship between Stress and Strain According to the Winding Angle of the Propulsion Shaft Applied with the Composite Material — 1,800 kN · m*

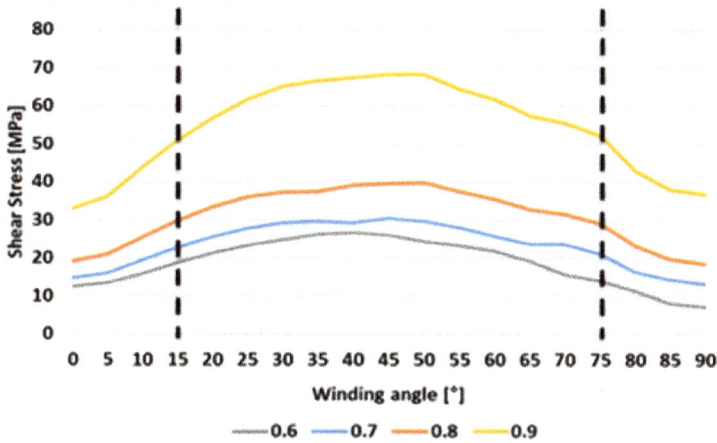

Fig. 8.13   Shear stress with the winding angle.

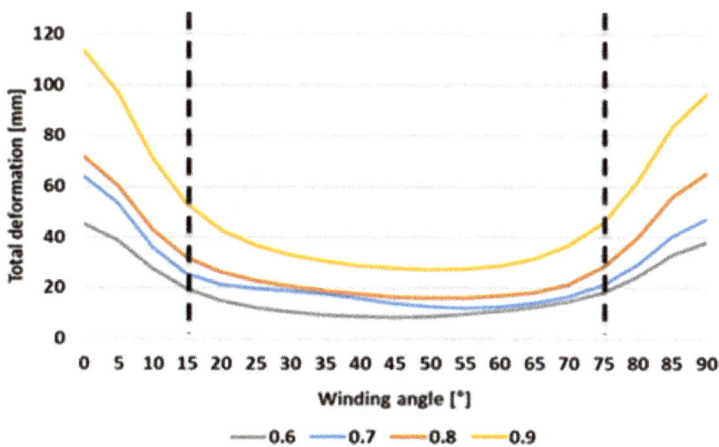

Fig. 8.14   Total deformation with the winding angle.

8.2.5.2 Relationship between Stress and Strain According to the Winding Angle of the Propulsion Shaft Applied with the Composite Material — 4,500 kN · m

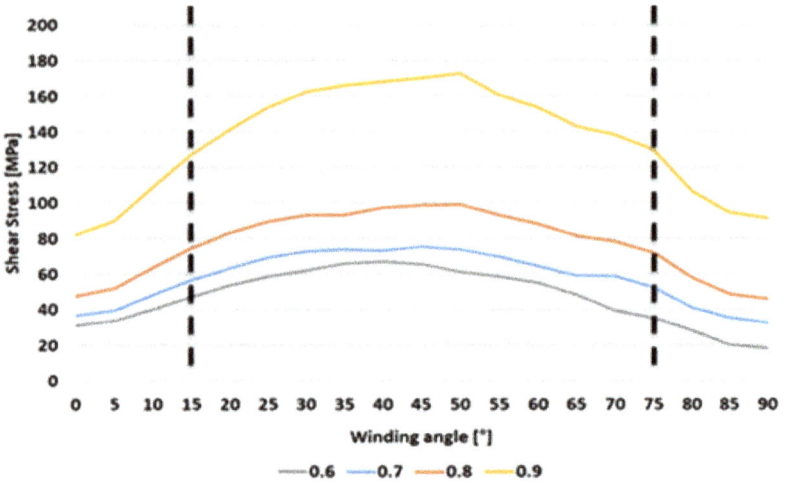

Fig. 8.15 Shear stress with the winding angle.

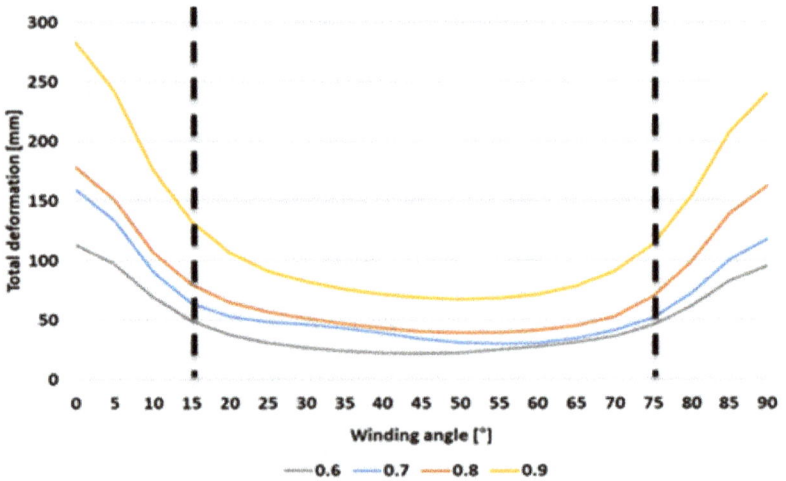

Fig. 8.16 Total deformation with the winding angle.

8.3 Torsional Test of Composites Tube

8.3.1 *Manufacturing of a Composite Material Tube by Filament Winding Method*

8.3.1.1 *Base Material (Matrix)*

The base material maintains the structure of the composite material by fixing the reinforcing material in the composite material, and serves to protect it from the external environment. For the base materials, metals, ceramics, carbon and polymers are used. The polymer system is divided into a thermosetting resin and a thermoplastic resin. In this study, epoxy resin, the most representative resin among the thermosetting resins, was used. As a study for application to propulsion shafts of ships, epoxy was applied to the propulsion shafts requiring high strength, because that epoxy is a resin with high chemical and corrosion resistance, and has superior mechanical properties, thermal properties, excellent adhesion to various materials, low shrinkage during curing, flexibility, excellent electrical properties and superior workability under various conditions than other base materials. KFR-120V epoxy resin manufactured by Kukdo Chemical was used as the base material, and KFH-141 was used as a curing agent. Table 8.4 shows the physical properties of the base material.

8.3.1.2 *Reinforcing Material (Reinforcement)*

Glass fiber, aramid fiber and carbon fiber are mainly used as the reinforcing materials for composite materials, and the general physical properties of each fiber are shown in Table 8.5. The industrial

Table 8.4 Properties of matrix (resin and hardener) at 25°C

Epoxy resin (KFR-120V)		Hardener (KFH-141)	
Equivalent weight (g/eq, EEW)	170~175	Total amine value (mgKOH/g)	500~700
Density (g/mL)	1.0~1.2	Density (g/mL)	0.8~1.0
Color (Gardner)	0.5 Max	Color (Gardner)	5 Max
Viscosity (cps)	800~1,100	Viscosity (cps)	5~50

Table 8.5 Properties of composite reinforcement

	Density (g/cm)^3	Tensile strength (GPa)	Modulus of elasticity (GPa)	Specific strength (cm)
Carbon fiber (T700-SC)	1.80	7.0	23	27.5
Duralumin steel (2024-T7)	2.77	0.4	74	1.6
E-glass fiber	2.55	3.4	74	13.7
Aramid fiber (Kevlar 49)	1.45	3.6	131	25.5
Stainless (SUS 304)	8.03	0.5	197	0.7
Basalt fiber (prince)	2.6	4.0	79.3	13.4

fibers have superior strength and elastic modulus than metallic materials such as steel, and have low specific strength with low density compared to metallic materials. For the case of carbon fiber particularly, the specific strength is 40 times that of steel and the inelasticity is about five times, so the physical properties are excellent compared to steel. Carbon fiber has the advantage of excellent physical/mechanical properties, but has low economic efficiency, so it has been applied to ultra-high-priced products such as aerospace, aviation and high-performance ships. Due to the increase in production efficiency, production volume and localization of materials, the price is lowered, and the application is expanding to other industries such as the automobile industry, shipbuilding industry and construction industry. In this study, T700-SC 24k carbon fiber manufactured by Toray was used.

8.3.2 Manufacturing Method for Composites Tube Specimen Using Filament Winding Process

In order to confirm the torsional strength of the composite material tube and the characteristics of the fastening portion, a specimen was prepared using a filament winding method. The filament winding process can be divided into four stages: mandrel preparation, pattern design, molding and processing and demolding. In the case of cylindrical mandrel, it is difficult to demold the composite material, so it must be molded after conducting release treatment sufficiently. After performing the release treatment, the design pattern is checked, the resin bath and the tension are adjusted for ensuring the constant resin contents of the fiber, and then the fiber is wound. After the winding process is completed, the extra resin on the surface is removed using a chisel. The additional winding for demolding is performed. According to the curing conditions of the resin, curing process is conducted using a rotary furnace. After demolding, the production is completed through the steps of quality control. The stages of the filament winding process and the laminating pattern by angle are shown in Figs. 8.17 and 8.18.

8.4 Materials and Model of the Composites Shaft

8.4.1 *Material Properties*

The mechanical properties of specimens were measured by tensile, compression and shear strength tests using a universal test machine according to the ASTM standards. More than 20 specimens were measured to ensure accuracy of the test data. The mechanical properties of the CFRP laminates for simulation are shown in Table 8.6.

8.4.2 *Analysis of Laminate*

Based on the results of basic properties test of carbon fiber composite material, analysis of the laminate was performed using the MCQ program. Table 8.7 shows the mechanical properties of CFRP laminates for simulation. The analysis of laminate was designed by dividing the mechanical properties of fiber and resin of CFRP. In this study, conclusions were reached through the simulation evaluation of the composite material, which is a lightweight ship material, in order to secure safety and operational efficiency. Reliability of the simulation results is secured through calculation of Fiber/Matrix/Ply calibration and laminate mechanics. After establishing the required mechanical properties of the hull, reinforcement material laminate

Fig. 8.17 Manufacturing process of filament winding; (a) Fixed mandrel, release treatment, (b) Adjust tension and resin bath, (c) Filament winding, (d) Remove excess resin and (e) Curing.

patterns that satisfy the required performance through reverse design should be implemented.

Fiber/Matrix/Ply Calibration: The ply calibration performance results show that the initial values of the mechanical properties of the carbon/epoxy composites in the E_{11} direction are higher than the

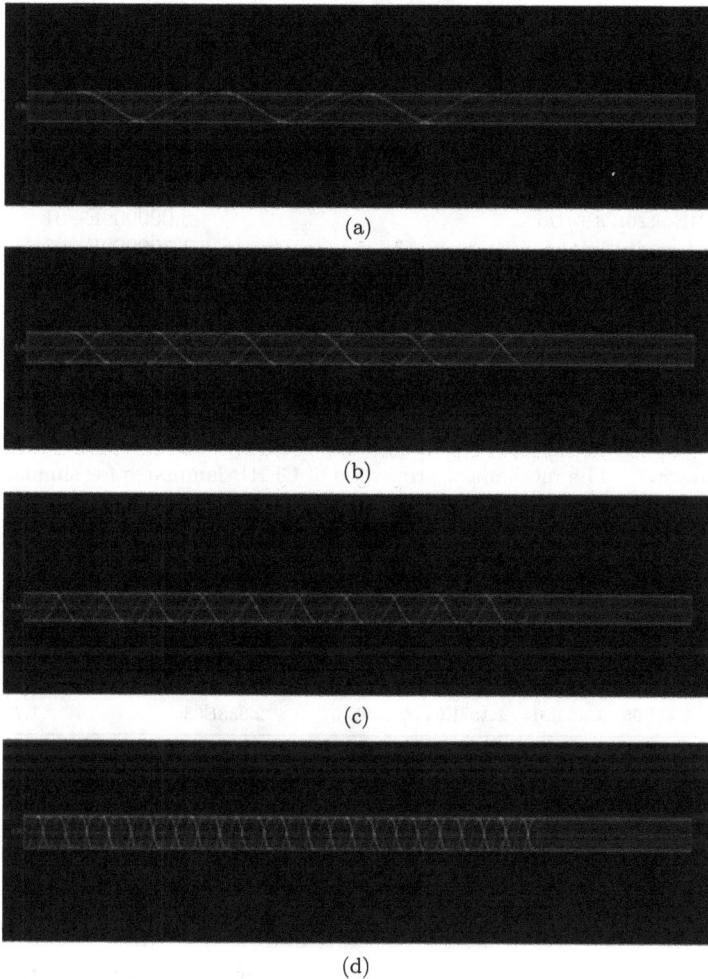

(a)

(b)

(c)

(d)

Fig. 8.18 Laminating pattern by angle; (a) winding angle 15°, (b) winding angle 30°, (c) winding angle 45° and (d) winding angle 75°.

calculated values, and the remaining values are the same. The ratio of reinforcing material to matrix material was 63:35.1 and the void content was calculated to be 1.9%. The result of the prediction of the fiber calibration performance showed that the initial value of the mechanical properties in the E11 direction of the carbon/epoxy is lower than the calculated value in the case of the carbon/epoxy, contrary to the prediction result of the Ply performance.

Table 8.6   The mechanical properties of CFRP laminates for simulation

Description	Unit	Initial value	COV
(E11)Longitudinal modulus	N/mm^2	1.350000E+05	
(E22)Transverse modulus	N/mm^2	1.000000E+04	
(G12)Shear modulus	N/mm^2	5.000000E+03	
(NU12)Poisson's ratio		3.000000E−01	
(S11T)Longitudinal tensile strength	N/mm^2	1.500000E+03	
(S11C)Longitudinal compressive strength	N/mm^2	1.200000E+03	
(S22T)Transverse tensile strength	N/mm^2	5.000000E+01	
(S22C)Transverse compressive strength	N/mm^2	2.500000E+02	
(S12S)In-plane shear strength	N/mm^2	7.000000E+01	

Table 8.7   The mechanical properties of CFRP laminates for simulation

	Modulus [N/mm^2]			Poisson	Strength [N/mm^2]				
Ply	E_{11}	E_{22}	G_{12}	NU12	S11T	S11C	S22T	S22C	S12S
	9.01E04	1E04	5E03	3E-01	1.5E03	1.2E03	15E01	2.5E02	7E01

	Modulus [N/mm^2]			Poisson	Strength [N/mm^2]		
Fiber	E_{11}	E_{22}	G_{12}	NU12	S11T		S11C
	1.43E05	1.432E04	1.107E04	2.198E-01	2.383E03		1.721E03

	Modulus [N/mm^2]	Poisson	Strength [N/mm^2]		
Matrix	E	NU	ST	SC	SS
	4.741E03	4.297E-01	7.528E01	3.764E02	1.091E02

Laminate Mechanics: Based on the calculated results of the Fiber/ Matrix/Ply Calibration, the composites were reconstructed to recalculate the physical properties of the Carbon/Epoxy composite materials. The tensile strength in the S11T direction was 1,515 MPa, which is almost the same as the initial value of 1,500 MPa. Figure 8.19 shows the performance results of Laminate Mechanics.

Progressive Failure: Eight ply reinforced materials were laminated in two directions [0°/90°/ + 45°/ − 45°] in order to determine the effect of load. Progressive failure was confirmed from the laminate design patterns, and damage prediction data for each stack direction were obtained. Figure 8.20 shows the damage data for Ply.

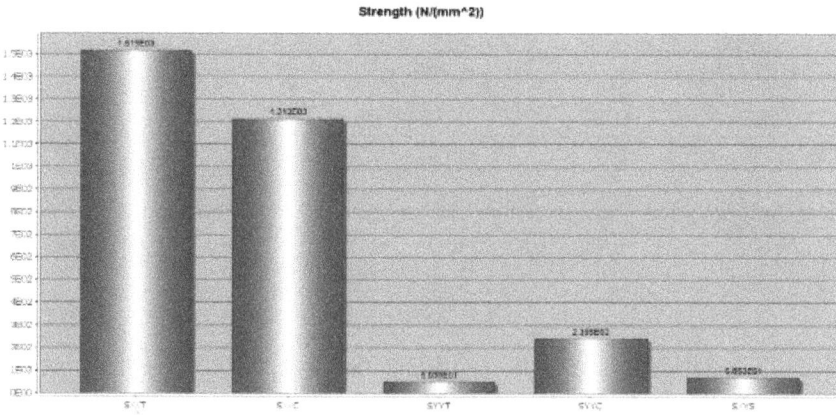

Fig. 8.19   Laminate mechanics.

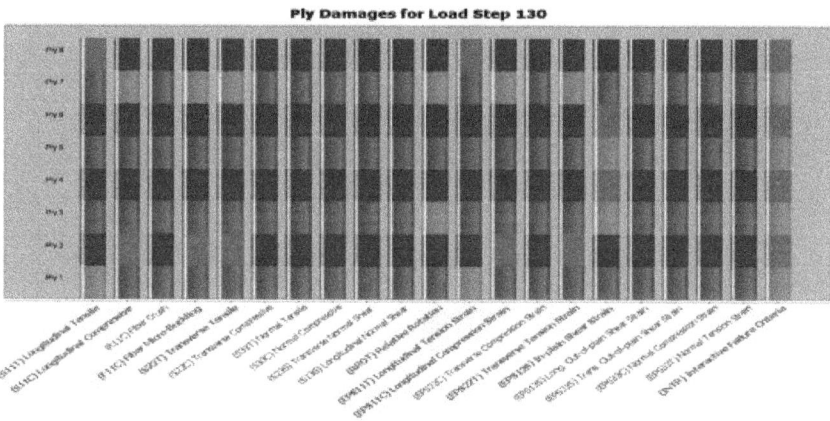

Fig. 8.20   Progressive failure.

Design Failure Envelope: This function was used to visually confirm the failure zone of Lamina or Laminates based on the failure criterion determined by the user when multiple loads were applied to the composite material. Figure 8.21 shows the composite failure criterion. The first quadrant represents the tensile strength, the third quadrant represents the compressive strength and the slope represents the shear strength. The yellow dot indicates the point at which the stress starts, and the red dot indicates the end of the stress.

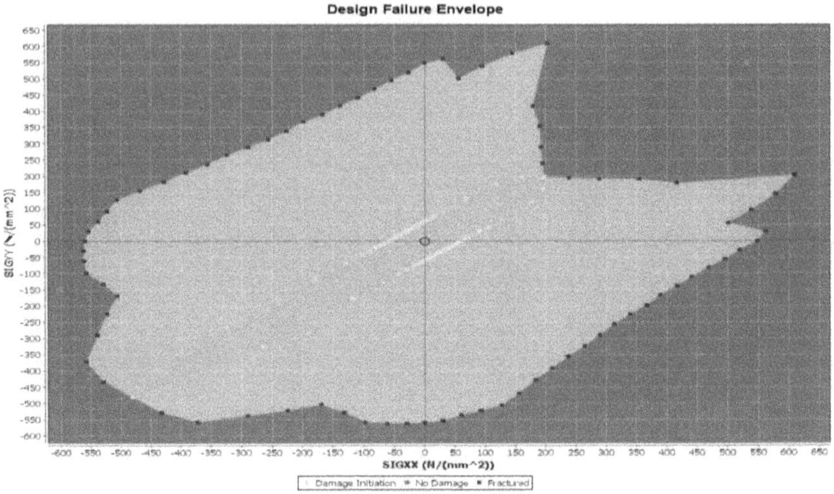

Fig. 8.21   Design failure envelope.

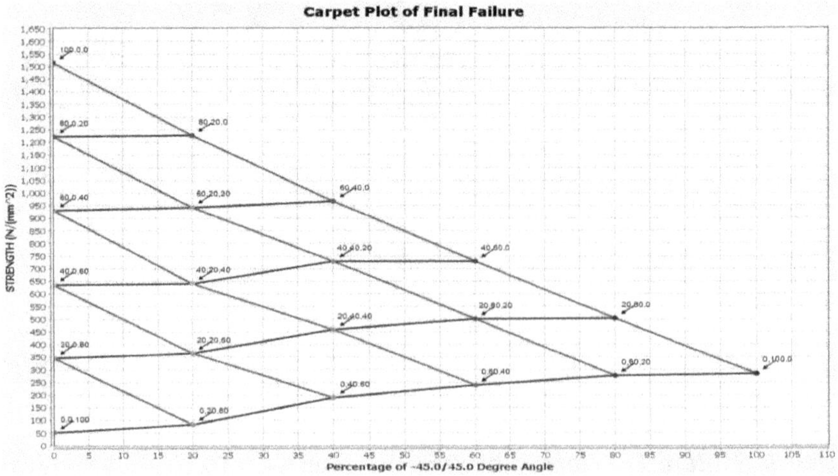

Fig. 8.22   Parametric carpet plots.

Parametric Carpet Plots: The results are shown in Fig. 8.22. Parametric Carpet Plots were performed to intuitively understand the mechanical properties of laminate and its tendency when the laminate contents were varied according to the angle of the ply laminated to the composites.

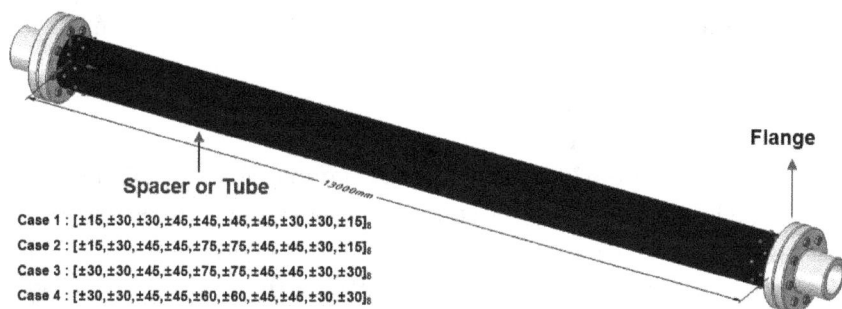

Fig. 8.23 The shape of intermediate shaft.

8.4.3 *The Finite Element Model*

In this study, finite element analysis was carried out using ANSYS ACP 19.0. The shaft of the ship to be developed is a hollow shaft with an outer diameter of 900 mm and a length of 13 m. The torque transmission capacity to be operated is 1,800 kNm. The shape of intermediate shaft is shown in Fig. 8.23. The layers and thickness required for the intermediate shaft were chosen by the shear stress equation of the hollow shaft. For the finite element model of the composite shaft, one end is fixed and torque is applied at the other end. The number of analysis cases according to the winding patterns of composites shaft are four cases. The analysis cases according to the stacking sequence of high-strength carbon/epoxy shaft are as follows:

Case 1: $[\pm15, \pm30, \pm30, \pm45, \pm45, \pm45, \pm45, \pm30, \pm30, \pm15]15$

Case 2: $[\pm15, \pm30, \pm45, \pm45, \pm75, \pm75, \pm45, \pm45, \pm30, \pm15]15$

Case 3: $[\pm30, \pm30, \pm45, \pm45, \pm75, \pm75, \pm45, \pm45, \pm30, \pm15]15$

Case 4: $[\pm30, \pm30, \pm45, \pm45, \pm60, \pm60, \pm45, \pm45, \pm30, \pm15]15$

8.5 Results and Discussions

8.5.1 *The Effect of Winding Angle*

In this study, the shear stress according to the diameter ratio was calculated from the shear stress equation of the hollow shaft. At a diameter ratio of 0.9, the shear stress was 36.6 MPa and the thickness of shaft was 45 mm. Also, in the effect of the shear modulus according to the laminated angle, the laminated angle of the fiber showed the best value at 45°. This means that to ensure structural safety in the design of the optimized shaft, the laminated angle of the fiber should be laminated at 45°. Meanwhile, as a result of the shear stress according to the winding angle, a graph was formed showing the shape of the parabola. At a diameter ratio of 0.9, the shear strength of 50MPa or more is required at a winding angle of 15°–75°.

8.5.2 *Torsion Experiment*

The most important property of a shaft in a power transmission system is to transmit the power without yielding. For this, the maximum torque of the shaft and the state of stress applied to the shaft should be understood. A state in which the shaft cannot work properly due to excessive deformation is called the breakage. As a specific form of the breakage, the damage in which materials are separated into two or more is the typical breakage. Even if the material is not destroyed, it is also considered to be a breakage when the deformation occurs excessively and the integrity of the component or its function cannot be performed. In the case of a composite material stacked structure, the case of damage is different for each layer, so more rigorous design is required than the metal materials.

The static torsion test is for evaluating the stability of the power transmission shaft against certain torque by applying a load to one end while the other end is fixed, or for confirming the mechanical properties such as torsional stiffness, yield strength and breaking strength by applying torque until breakage using the relationship between torque and torsion angle. Therefore, through experiments, it is intended to derive the most stable stacking patterns for winding stacking angles of 15°, 30°, 45° and 75°, and apply the results to large structures.

Five specimens each with length of 280 mm, an outer diameter of 22 mm and a thickness of 1 mm were prepared by a filament winding

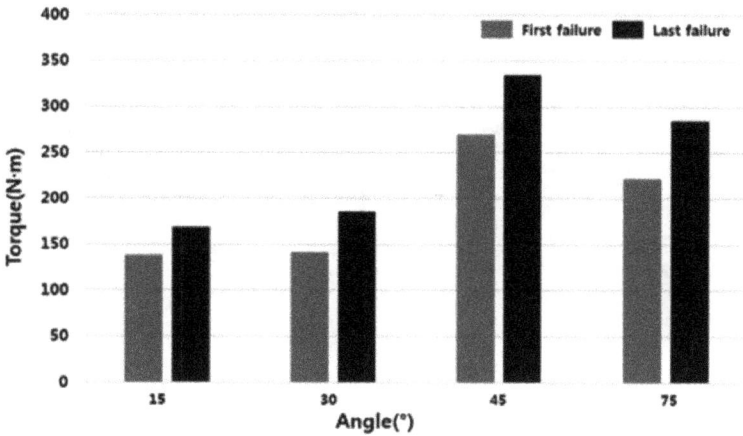

Fig. 8.24 Comparison of torsional test average values by angle.

process for each stacking angle, and the metal jig was bonded using an adhesive. The strain gauge was installed at the center where the specimen fracture was expected to be tested.

Figures 8.24 and 8.25 show the torque value according to the winding angle of the composite material. According to the graph of the torsion test, the torque decreases primarily, and after the additional decrease, the torque reaches the maximum value and the breakage occurs. For the maximum torque, it was $169\,\mathrm{N}\cdot\mathrm{m}$ in average at a winding angle of $15°$, $186\,\mathrm{N}\cdot\mathrm{m}$ in average at a winding angle of $30°$, $335\,\mathrm{N}\cdot\mathrm{m}$ in average at a winding angle of $45°$ and $285\,\mathrm{N}\cdot\mathrm{m}$ in average at a winding angle of $75°$. However, it is estimated that the value when the first torque is reduced than the maximum torque value is important, because the composite material has the performance and stability of the structure according to the breakage of each layer. As for the torque when the first breakage occurs according to angles, it was $138\,\mathrm{N}\cdot\mathrm{m}$ in average at a winding angle of $15°$, $141\,\mathrm{N\cdot m}$ in average at a winding angle of $30°$, $269\,\mathrm{N\cdot m}$ in average at a winding angle of $45°$ and $221\,\mathrm{N\cdot m}$ in average at a winding angle of $75°$. At the points where the first breakage occurred and the last breakage occurred, that is, when the torque had the maximum value, the torque value was measured as the highest value when the

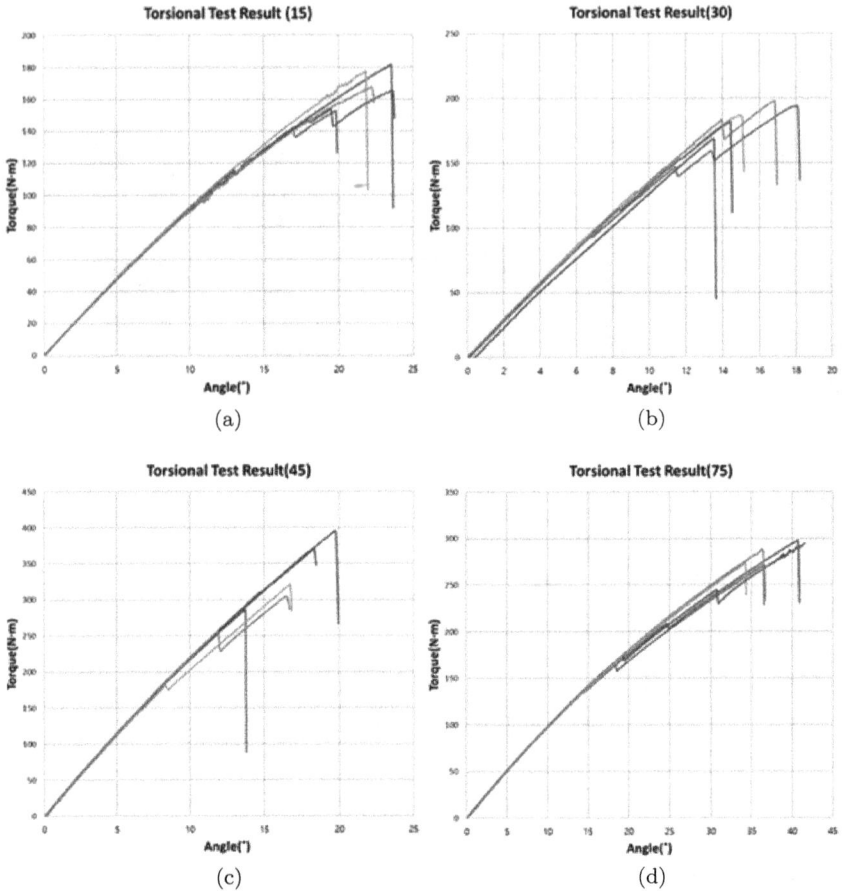

Fig. 8.25 Torsional test results according to the winding angles (a) 15°, (b) 30°, (c) 45° and (d) 75°.

winding angle was 45° and then the torque values were decreased in order of the winding angle of 75°, 30° and 15°.

This experiment is a coupon test based on a specimen. In this case of a propulsion shaft that is actually applied, there are many variables to be considered depending on length, diameter, etc., so it is impossible to determine the winding stacking pattern only by the factor of the torsion test. It is predicted that the winding angle will increase as the size of the propulsion shaft increases, and it is estimated that the optimum torque value will be shown in a stacking pattern of 45° or more.

8.5.3 *The Result of Tsai–Wu Failure Criteria*

In order to verify structural safety of the composites intermediate shaft, it was verified using Tsai–Wu failure criteria. A verification of structural stability was made through finite element analysis. The result of finite element analysis is shown in Figs. 8.26 and 8.27. The review showed that the Inverse Reserve Factor (IRF) was less than 1 in all 4 cases, but Case 2 was the most stable. This means that for composite stresses delivered to the structure, a winding angle of 15° and 75° is required.

8.5.4 *The Modal Analysis Results*

The mode analysis was performed on four stacking cases and confirmed the natural frequency and 6th mode shape. In addition, a fixed boundary condition was established assuming that one end of the shaft is coupled with the flange coupling.

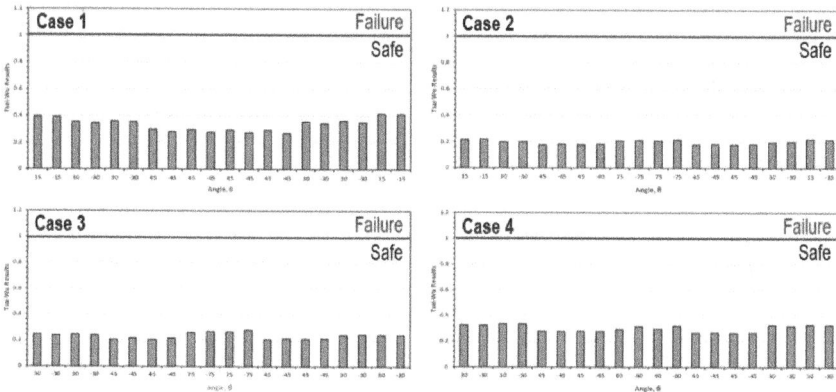

Fig. 8.26 The result of Tsai–Wu failure criteria.

No.	Angle	Case 1		Angle	Case 2		Angle	Case 3		Angle	Case 4	
		Min	Max		Min	Max		Min	Max		Min	Max
1	15	0.056059	0.39447	15	0.061974	0.21595	30	0.049161	0.24656	30	0.043083	0.32571
2	-15	0.056903	0.39373	-15	0.063093	0.21933	-30	0.050464	0.23728	-30	0.045484	0.32522
3	30	0.047907	0.35486	30	0.055333	0.19723	30	0.049179	0.24448	30	0.04357	0.33606
4	-30	0.050035	0.34841	-30	0.05626	0.20059	-30	0.050436	0.24006	-30	0.045473	0.33533
5	30	0.048174	0.3614	45	0.052412	0.17791	45	0.045698	0.20671	45	0.039587	0.27915
6	-30	0.050077	0.35463	-45	0.054019	0.18257	-45	0.048154	0.21637	-45	0.043551	0.27766
7	45	0.044724	0.29809	45	0.052494	0.17881	45	0.045817	0.20799	45	0.039695	0.28016
8	-45	0.048083	0.28294	-45	0.054255	0.18281	-45	0.048472	0.21766	-45	0.043699	0.27747
9	45	0.045228	0.29644	75	0.061432	0.20909	75	0.054747	0.25925	60	0.042503	0.29616
10	-45	0.048274	0.27953	-75	0.066856	0.21036	-75	0.060756	0.2675	-60	0.045559	0.3157
11	45	0.045305	0.29473	75	0.061466	0.20872	75	0.054629	0.26377	60	0.042756	0.29891
12	-45	0.048497	0.27636	-75	0.067074	0.21755	-75	0.060915	0.28023	-60	0.046016	0.31889
13	45	0.045322	0.29293	45	0.052979	0.17978	45	0.046192	0.20972	45	0.040217	0.26588
14	-45	0.048732	0.2734	-45	0.05471	0.18182	-45	0.049041	0.2135	-45	0.044293	0.26986
15	30	0.048749	0.35681	45	0.05295	0.18057	45	0.04621	0.21075	45	0.040393	0.26692
16	-30	0.050852	0.3469	-45	0.054943	0.18225	-45	0.049257	0.21359	-45	0.044492	0.26962
17	30	0.048904	0.36246	30	0.055908	0.19596	30	0.050042	0.24012	30	0.044499	0.32717
18	-30	0.050849	0.35237	-30	0.057216	0.20102	-30	0.051307	0.24423	-30	0.046333	0.3182
19	15	0.056647	0.41599	15	0.063396	0.21952	30	0.050105	0.24066	30	0.044897	0.32992
20	-15	0.057523	0.41473	-15	0.063888	0.21666	-30	0.051203	0.24218	-30	0.046208	0.32684

Fig. 8.27   The result of FEA.

The 6th mode shapes of Case 2 having the highest natural frequency among the four cases are shown in Fig. 8.28. As shown in the figure, the first bending mode occurred in (a)–(d). In addition, the breathing mode shape can be seen in (e) and the second bending mode occurs in (f). Typically, if the rotating shaft meets the critical speed corresponding to the first bending mode, the rotational vibration greatly increases.

Table 8.8 shows that Cases 2 and 3, which contain layers stacked as ±75 degrees, have higher natural frequency than Cases 1 and 4. This result is considered more secure because Cases 2 and 3 meet the critical speed at higher natural frequencies. Typically, higher order modes have higher natural frequencies, and the results in Table 8.8 show the same tendency.

8.5.5 Campbell's Diagram Results

The effect of critical speed according to the stacking sequence is shown in Fig. 8.29. Campbell's diagram results in Fig. 8.29 consider

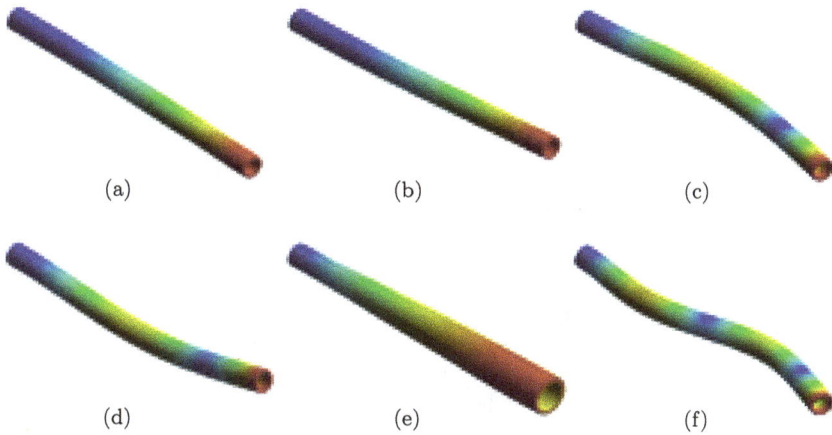

Fig. 8.28 6th Mode shapes of Case 2 (a) 1st, (b) 2nd, (c) 3rd, (d) 4th, (e) 5th and (f) 6th.

Table 8.8 The natural frequency results according to the stacking sequence and pattern

Mode	Case 1	Case 2	Case 3	Case 4
1st	3.1828	4.4496	4.3421	2.7607
2nd	4.0691	4.812	4.7093	4.3555
3rd	19.107	26.758	26.098	22.568
4th	23.137	28.173	27.32	24.735
5th	51.641	63.428	64.524	60.866
6th	61.163	71.549	70.007	64.644

a 15%–20% margin between operating and critical speeds used in the industry. The critical speed is highest in (b) of Fig. 8.29. Also, (b)–(d) of Fig. 8.29 exceeded the range of critical speed margins. This indicates that the operating speed of the shaft is safe for the first natural frequency.

S.-W. Yoon

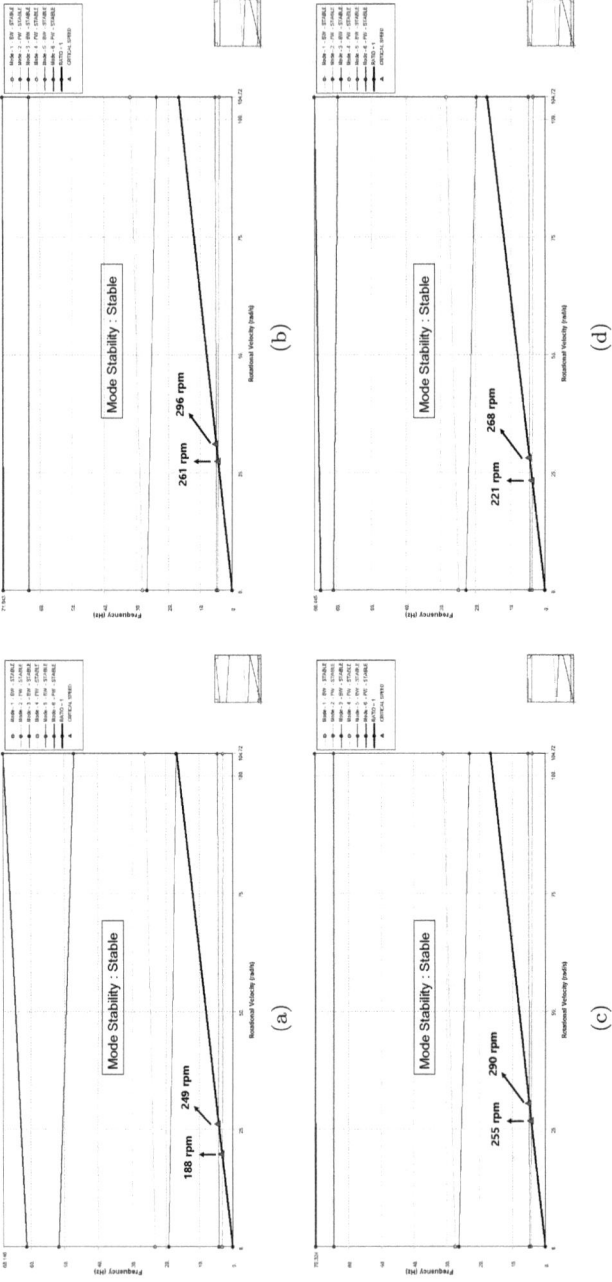

Fig. 8.29 Campbell's diagram results (a) Case 1, (b) Case 2, (c) Case 3 and (d) Case 4.

8.6 Conclusions

8.6.1 *The Analysis of Laminate*

In this study, conclusions were reached through the simulation evaluation of the composite material, which is a lightweight hull material, in order to secure safety and operational efficiency. Reliability of the simulation results is secured through calculation of Fiber/Matrix/ Ply calibration and laminate mechanics. The ply calibration performance results show that the initial values of the mechanical properties of the CFRP in the E_{11} direction are higher than the calculated values. The laminate mechanics results show that the tensile strength in the S11T direction was 1,515 MPa. After establishing the required mechanical properties of the hull, reinforcement material laminate patterns that satisfy the required performance through reverse design should be implemented.

8.6.2 *The Finite Element Analysis*

In this study, the possibility of the application of composite materials for a marine propulsion shaft is analyzed by FEA. The structural safety of the propulsion shaft with composite material is not significantly superior to that of the propulsion shaft with stainless steel material. However, considering the benefits of weight reduction and the reduction of the number of components, the possibility of replacing propulsion shafts with composite materials will be sufficient. The main conclusions of this study are summarized as follows:

- In this work, the winding angle and the order of layers of the composites intermediate shaft were optimized by the FEA.
- The winding angle has a great effect on the structural strength of the composites intermediate shaft. The optimum fiber orientation for maximum torsional strength should be close to 45°. In addition, it is necessary to place a winding angle of 15° and 75° to prepare for the combined stresses transmitted to the structure.
- The optimal order of winding for maximum torque is Case 2. Based on this analysis, a shaft design technique was proposed.

8.6.3 *The Effect of Winding Angle*

In this study, the intermediate shafts made up of T700-SC multi-layered composites have been designed to replace the steel shaft of a ship. The major conclusions in this study are summarized as follows:

- The designed CFRP intermediate shafts are optimized by using FEA for stacking sequence, torque transmission capacity and bending vibration characteristics.
- The fiber orientation angle has a great influence on the natural frequency of the shaft. The carbon fiber should be closely oriented at $\pm 30°$ and $\pm 75°$ to improve the modulus of elasticity in the direction of length of the shaft and to increase the natural frequency. And, the optimum fiber orientation for maximum torsional strength is $\pm 45°$.
- The optimum stacking sequence that can be operated at the target rotational speed is Case 2, and based on these analysis results, we have proposed guidelines that can be applied to the intermediate shaft design.

Acknowledgments

This research was a part of the project titled "Development of Fire Resistant, High-Pressure and High-Strength Composite Material to Replace Green-ship Steel Outfit" (Projects No. 20160271), funded by the Ministry of Oceans and Fisheries, Korea.

References

[1] The Japan Machinery Federation (JMF) and the Next Generation Metals and Composites Research and Development Association (GMMR), 2008 Investigation Report on the Application of Thermoplastic Composites to Aircraft Field, March 2009.

[2] Matsunaga *et al.*, Development of Low-Cost CFRTP Processing Technology (Report 1), CFRTP Analysis Technology and the Effect of Molding Conditions on Mechanical Properties, Research Report of Seibu Industrial Technology Center, Hiroshima Prefectural Institute of Technology, No. 54, 2011, pp. 1–4.

[3] NEDO News Release, NEDO Technical Innovation Award to Carbon Fiber Composite Materials Development Project, http://www.nedo.go.jp/news/ press/AA5_100130.html.

[4] A. Takahashi, Special Issue on Automotive Materials: Evolution and Challenges of Materials — Various Possibilities and Challenges of Automotive Materials from the Viewpoint of Recycling, Japan Automobile Manufacturers Association, Inc., http://www.jama.or.jp/lib/jamagazine/200603/05. html.

[5] A. Kitano, *J. Chem. Educ.*, 59(4), 2011, 226–229.

[6] I. De Baere, W. Yan Paepegem, C. Hochard and J. Degriek, *Polym. Test.*, 30, 2011, 663–672.

[7] D. Backe and F. Balle, *Compos. Sci. Technol.*, 126, 2016, 115–121.

[8] Y. Zhou and P. K. Mallick, *Polym. Compos.*, 27, 2006, 230–237.

[9] V. Bellenger, A. Tcharkhtchi and P. Castaing, *Int. J. Fatigue*, 28, 2006, 1348–1352.

[10] R.-I. Murakami, M. Masuda and T. Nonomura, *Jpn. Soc. Mech. Eng.*, 58(545), 1992, 9–14.

[11] Y. Uematsu, T. Kitamura and R. Ohtani, *Compos. Sci. Technol.*, 53, 1995, 333–341.

[12] K. Tanaka, K. Oharada, D. Yamada and K. Shimizu, *Procedia Structural Integrity*, 2, 2016, 058–065.

[13] C. M. Socino and E. Moosbrugger, *Int. J. Fatigue*, 30, 2008, 1279–1288.

[14] S. Mortazavian and A. Fatemi, *Adv. Mater. Res.*, 891/892, 2014, 1403–1409.

[15] R.-I. Murakami, Y. Kim and K. Kusukawa, *Fundamental Strength and Fracture of Materials*, Fikuro Publishing, 2009, p. 74.

[16] A. K. Haldar and S. Senthilvelan, *Key Eng. Mater.*, 471, 2011, 173–178.

[17] A. Fajrin, W. Solafide and R. Murakami, The Effect of Fiber Ply Orientation, Laminate Layer and PLA Content on Mechanical Properties of Carbon Fiber Reinforced Bio Plastic Composites, 2nd International Conference on Nanomaterials and Advanced Composites, August, 2019, Taipei, Taiwan.

[18] J. Horst and J. L. Spoormaker, *Polym. Eng. Sci.*, 6(36), 1996, 2718–2726.

[19] K. Tanaka, T. Kitano and N. Egami, *Eng. Fract. Mech.*, 123, 2014, 44–58.

[20] M. F. Arif, N. Saintier, F. Meraghni, J. Fitoussi, Y. Chemisky and G. M. Robert, *Compos. Part B, Eng.*, 62, 2014, 55–65.

[21] A. Bernasconi, P. Davoli, A. Basile and A. A. Fillipi, *Int. J. Fatigue*, 29, 2007, 199–208.

[22] R. Selzer and K. Friedrich, *Compos. Part A*, 28A, 1997, 595–604.

[23] D. Flore, K. Wegener, D. Seel, C. C. Oetting and T. Bublat, *Compos. Part A*, 90, 2016, 359–370.

[24] K. Tanaka, K. Oharada, D. Yamada and K. Shimizu, *Procedia Structural Integrity*, 2, 2016, 058–065.

Chapter 9

Automotive and Elevator Composite Structures

Sung-Youl Bae

Technology Convergence Division, Korea Institute
of Ceramic Engineering & Technology,
Republic of Korea
bsy@kicet.re.kr

This chapter deals with design, analysis and test of advanced composites for automotive and lift applications. The chapter covers development examples of composite structures replacing existing steel parts in automotive and elevator structures. In addition, the chapter focuses on the procedures and methods of design and evaluation of automotive and lift composite components.

9.1 Introduction

According to the regulations of vehicle fuel consumption, weight reduction of automotive components is essential. In order to improve the fuel efficiency of vehicles, the development of automotive parts using advanced materials, such as high-strength steel, aluminum and carbon fiber reinforced plastics (CFRP), is actively underway. Now, high-strength steel and non-ferrous metals are mainly applied to reduce the weight of automotive components. In addition, CFRP will be applied on automotive components in the near future because of its various merits as compared to metals. Currently, the development of composite components by use of CFRP, which is lightweight, strong and durable, is in progress. However, it has not been easy to commercialize due to its high price. Some researchers say that

CFRP could be applied on automotive components when the cost of CFRP is 25% of the current cost. Therefore, it is essential to develop technology to reduce the cost of CFRP through the application of automated processes and the development of low-cost carbon fibers.

The necessity for the weight reduction of elevator components has been raised. Moreover, developments of lightweight components are actively conducted by various elevator system manufacturers because of their various benefits, such as energy savings, installation merits, easier maintenance, repair efficiency and new elevators types' emergence.[1-3] The weight of the existing elevator components is heavy because steel is applied to most components of the existing elevator cabins and suspension ropes, among others. For this reason, difficulties occur in the installation, maintenance and repair of elevators. Recently, the development of elevator components by means of composite materials has been actively undertaken by elevator system manufacturers to improve the performance of elevator systems. Developments of lightweight components for elevator cabins, elevator frames, suspension ropes and sheaves have been conducted by some elevator manufacturers.

9.2 Composite Applications in Automotives and Elevator

Nowadays, the development of lightweight technology for automotive components is being emphasized because of environmental regulations and eco-friendly technology trends around the world. There are certain methods to improve the fuel efficiency of transportation machinery, such as automobiles and elevators. The methods include the improvement of driving parts, advancement of aerodynamic performance and weight reduction of the components. Currently, the automotive components have already been developed to minimize fuel efficiency through many optimizations in the field of driving parts and aerodynamic design. Therefore, the weight reductions of automotive parts, covering the development of structures, construction methods and materials continue to be the focus of numerous automotive manufacturers.

Damped aluminum laminate (DAL) panels have been widely used for automotive materials because of their lightweight, low-vibration and low-noise characteristics compared to conventional steel materials. Nowadays, damped metal panels are applied to on board charger (OBC) covers for electric vehicles to improve the noise, vibration, harshness (NVH) performance of the vehicles. The damped metal laminate is composed of two thin metal skins with a viscoelastic resin of about 20–80 μm, with these two sheets bonded together. When the aluminum hybrid panel is applied to an automobile material, the automotive components are significantly lightened compared to conventional steel components.[4] Moreover, the riding quality is improved because vibration energy is absorbed by polymer resin which is applied in the hybrid panel. For these reasons, the applications of DML materials to the components of automobiles and elevators have widely expanded.

The development of CFRP automotive components are actively conducted although there are not many cases of CFRP automotive component commercializations due to the high cost of CFRP components. Now, the development of materials, parts and equipment for application to automotive components is progressing at the same time. The development of CFRP boom structure for concrete pump truck (CPT) is a good example of an automotive CFRP component. CPT is a construction machine that transports concrete at construction sites. The weight of the CPT boom and the cylinder corresponds to about 55% of the total CPT weight. It is necessary to lighten the CPT boom by reducing the weight of the vehicle in order to save energy. The existing steel CPT end boom is replaced with CFRP, which is 30% lighter than the existing components and which reduces the weight of the car body, the load on the operating part and the maintenance and repair costs while improving fuel efficiency.

The development of ultra-lightweight elevators is the key technology for linear-driven rope-less elevator system, while the development of ultra-lightweight elevators by CFRP is ongoing in some companies. The rope-less elevator system is a scheme in which several elevators can be operated in a single hoistway, and the elevator can move in the vertical/horizontal direction to maximize the transport capacity of high-rise buildings. In order to operate this system, it is necessary to reduce the weight of the elevator by more than 50% compared with the existing steel elevators. The ultra-lightweight elevator has been developed by applying the composite material and design optimization of the frame by ThyssenKrupp. In addition, low vibration elevators have been developed by applying damped aluminum laminates to elevator car walls to reduce weight and to improve NVH performance among elevator manufacturers. DAL is a composite material in which two aluminum sheets measuring 0.7–1.0 mm are bonded together by viscoelastic resin. Since vibration energy is absorbed by the resin, the structure to which the material is applied is characterized by reduced noise and vibration. Therefore, when the material is applied to the elevator components, such as car wall, ceiling and platform, it is expected to reduce the weight of the elevator parts and improve the rider's comfort.

Elevator suspension belts by CFRP have been developed by some elevator manufacturers because conventional steel wire rope is difficult to apply in high-rise buildings due to the risk of failure caused by the heavy weight of the steel wire rope. CFRP rope has several advantages such as being lightweight, strong and durable compared to the existing steel wire rope. Moreover, small-sized sheave could be applied to an elevator system by means of a CFRP belt because of its flexibility as compared to steel rope.

9.3 Design of Automotive and Lift Structures by Composites

A typical design process of composite parts in the automobile and elevator is shown in Fig. 9.1. The aerodynamic design and the structural design of the composite parts are performed simultaneously as shown in the figure. In addition, the designs of composite parts are completed when the performances satisfy the specific standards or guidelines.

The newly proposed design procedure of composite structures is shown in Fig. 9.2. As shown in the figure, the integrated system load analysis is applied on the traditional design procedure, thus requiring

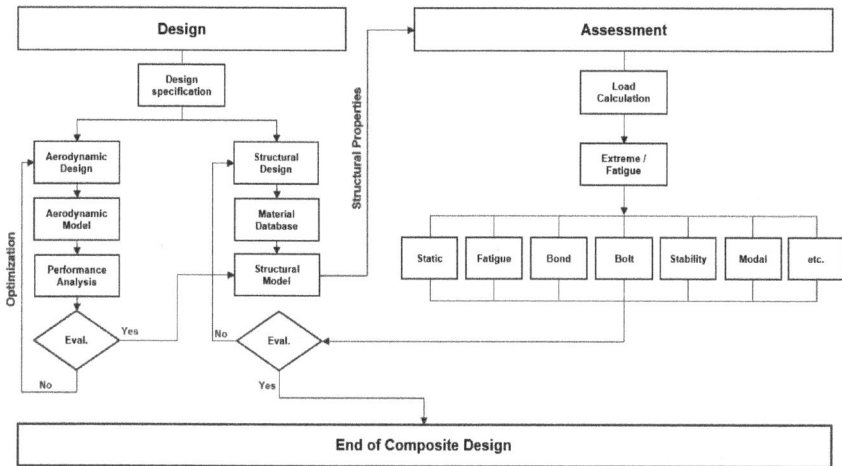

Fig. 9.1 Design procedure of composite components.

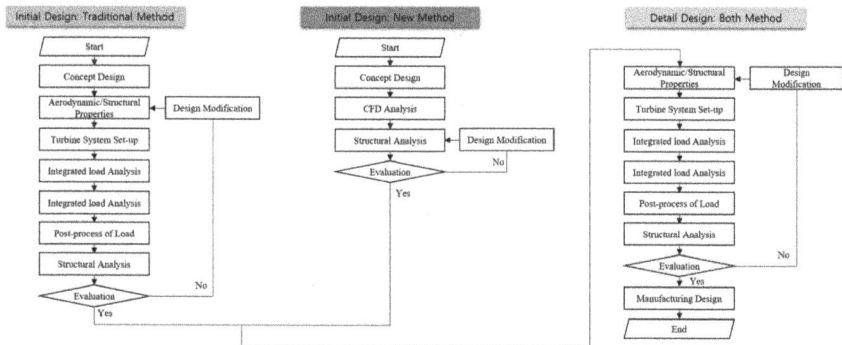

Fig. 9.2 New design procedure of composite structures.[6]

a complex process for calculating the aerodynamic and structural characteristics and the setup of a system for load analysis, with significant time for load analysis and post-process.[5] On the other hand, unidirectional fluid-structure interaction analysis is applied on the new design procedure to calculate the surface pressure of the structures. The pressure load calculated using CFD is applied on the structural evaluation. By using the new method, strain and stress on the structure may be calculated by applying the CFD pressure to

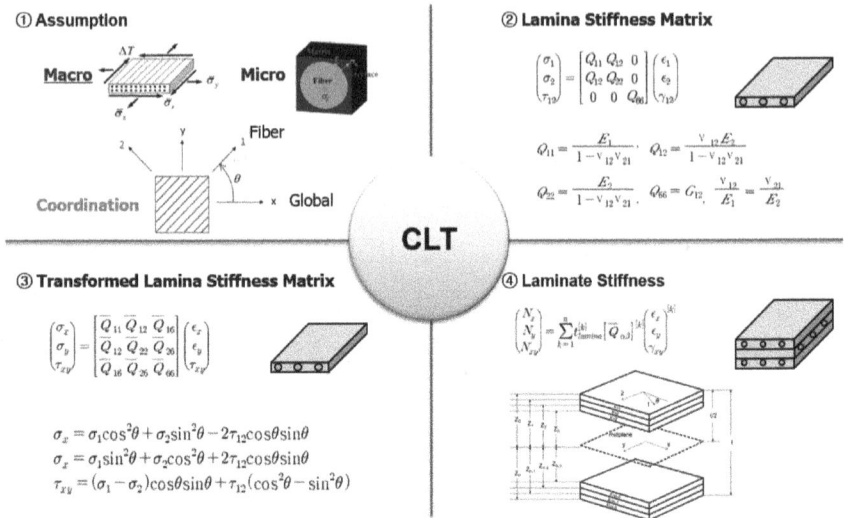

Fig. 9.3 Composite design by composite lamination theory.[7-9]

the newly designed model without performing additional procedures even in the case of a redesign in the preliminary design, thus being simpler than the conventional approach.

The design process of composite materials is shown in Fig. 9.3. Structural mechanics of composite materials are divided into macro-mechanics and micro-mechanics. Fibers, resins and their interfaces are considered separately in the micro-mechanics of composites. Moreover, the combined properties of each direction of the fiber and resin are considered in the macro-mechanics of composites. The equivalent stiffness by composite lamination theory should be calculated because the composite material is composed with different kinds of materials and laminar directions. The coordinate system is divided into fiber coordinate system and the global coordinate system. First, the stiffness for each laminar is obtained by some equations of composite mechanics in which the stiffness matrix could be derived. Moreover, equivalent stiffness of the laminate could be calculated by integrating all the laminar in the composite laminate.

The criteria for the structural integrity evaluation of composite materials are shown in Fig. 9.4. Generally, composite structures

Contents	Mode			Criterion
Laminate Failure: Fiber Failure	Tension		Compression	$\dfrac{\sigma_1}{X_t} = 1$ for $\sigma_1 > 0$ or $\dfrac{\sigma_1}{X_c} = 1$ for $\sigma_1 < 0$
Laminate Failure: Inter Fiber Failure	Tension (Mode A)	Shear (Mode B)	Compression (Mode C)	Mode A, B, C criteria
Sandwich Failure: Delamination		Out of Plane Shear		Finite Element Analysis: Solid Model+Cohesive Element
Sandwich Failure: Core Shear		Out of Plane Shear		· Coordination · Failure criterion $f = \dfrac{\tau_{13}}{S_{XZ}} + \dfrac{\tau_{23}}{S_{YZ}}$
Sandwich Failure: Face Sheet Wrinkling		Compression		· Coordination · Allowable stress $\sigma_{crinkle} = -Q(E_f E_c G_c)^{1}$ · Failure criterion $f = \dfrac{\sigma_j}{\sigma_{crinkle,j}} + \left(\dfrac{\sigma_k}{\sigma_{crinkle,k}}\right)^{2}$

Fig. 9.4 Failure criteria of composite evaluations.[10–12]

should be evaluated for fiber failure, interfiber failure, delamination and core failure. Each failure criterion is developed based on various test results, and the failure assessment of the composite structure should be performed by applying each failure criterion.

9.4 Development Examples of Automotive and Lift Structures

9.4.1 Development of Automotive Components by Aluminum Composites

OBC cover for electrical vehicle has been developed by aluminum laminate panel because of vehicle weight reduction and NVH performance improvement. The characteristics were evaluated to verify performances for commercialization by finite element analysis (FEA) and material tests. The newly designed component could be reduced by more than 60% compared to the existing component as shown in Table 9.1. Moreover, the component has similar NVH performance compared with the existing product. In order to verify the results of the adhesive strength analysis between the existing product and the

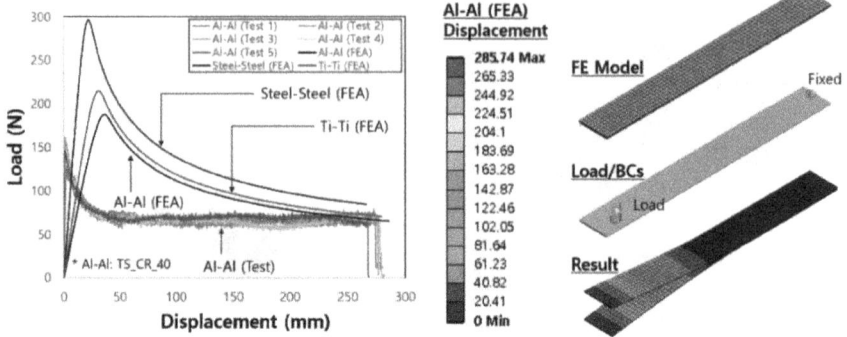

Fig. 9.5 Load-displacement results of T-peel for designed material by tests and FEA.[13]

Table 9.1 Comparison of design specification between the steel component and aluminum part for OBC cover[13]

Contents	Material	Weight (kg)	Thickness (mm)	Damping ratio (%)	Adhesion strength (N/25 mm)
Steel OBC Cover	Damped Steel Laminate	1.59	1.6	2.5	100
Aluminum OBC Cover	Damped Aluminum Laminate	0.54	1.6	2.4–2.6	65–80

developed component, the peel strength was verified via FEA. The results between the structural analysis and the adhesion strength tests were relatively similar, while the DAL had a lower adhesive strength than the conventional damped steel laminate as shown in Fig. 9.5. Table 9.1 shows the material design specifications to be analyzed in this study. Various materials were fabricated for the DAL according to the resin type, resin thickness and surface treatment material as illustrated in Table 9.2. Test equipment, test methods and test specimens are shown in Fig. 9.6. Also, the adhesion strength of various kinds of specimens is shown in Fig. 9.7. Mechanical tests and FEA simulations were conducted to determine the optimal design specifications of aluminum composites for automotive components.

Table 9.2 Test specimens of damped aluminum laminates[13]

No.	Name	Skin material (thickness)	Resin material	Resin thickness	Surface treatment material
1	TS-Cr_40	Aluminum (0.8 mm + 0.8 mm)	Polyurethane (Thermosetting)	40 μm	Cr
2	TS-Cr_30	Aluminum (0.8 mm + 0.8 mm)	Polyurethane (Thermosetting)	30 μm	Cr
3	TS-Cr_20	Aluminum (0.8 mm + 0.8 mm)	Polyurethane (Thermosetting)	20 μm	Cr
4	TS-TiZr_40	Aluminum (0.8 mm + 0.8 mm)	Polyurethane (Thermosetting)	40 μm	TiZr
5	TP-Cr_40	Aluminum (0.8 mm + 0.8 mm)	Isobutylene Rubber (Thermoplastic)	40 μm	Cr
6	TP-TiZr_40	Aluminum (0.8 mm + 0.8 mm)	Isobutylene Rubber (Thermoplastic)	40 μm	TiZr

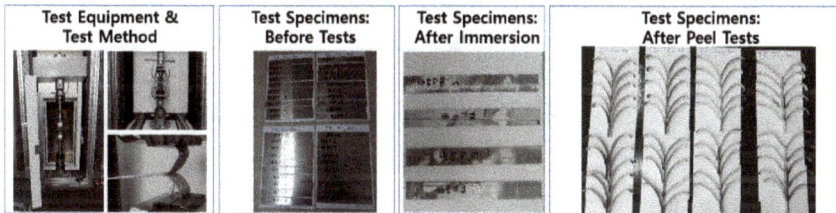

Fig. 9.6 Test equipment, method and specimens.[13]

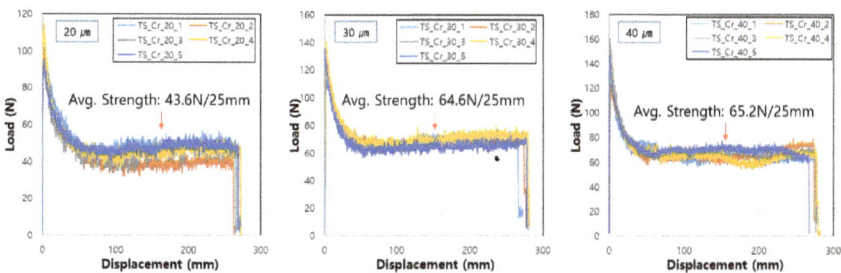

Fig. 9.7 Comparison of adhesive strength behavior according to adhesive resin thickness.[13]

A difference was confirmed in the peeling strength depending on the resin thickness, surface treatment agent and resin type.

9.4.2 Development of Automotive Components by CFRP

The lightweight boom structure for CPT was developed by CFRP (Fig. 9.8). Comparison of design specification between existing structures and development is shown in Table 9.3. Around 32% weight reduction of end boom structure was possible by using CFRP compared with the existing component. The maximum thickness of the existing structure was 18 mm while the CFRP boom was designed to have a maximum thickness of 36 mm.

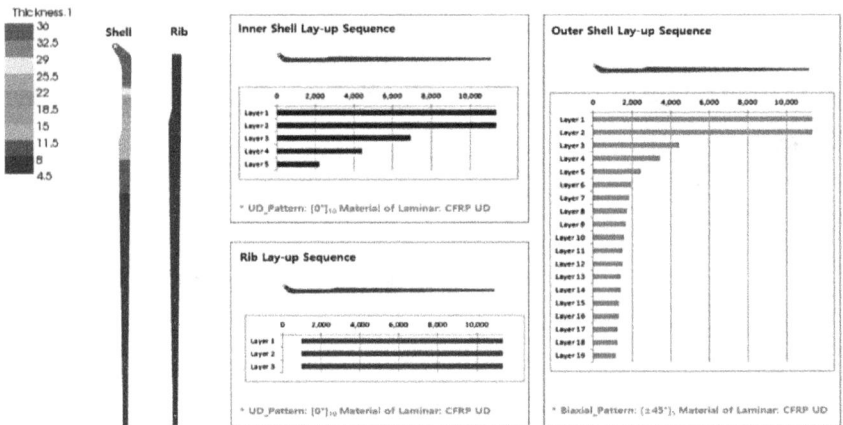

Fig. 9.8 Lay-up sequence and thickness plot of CFRP CPT end boom.[14]

Table 9.3 Comparison of design specification between original design and new design[14]

Contents	Material	Thickness (mm)	Weight (kg)	Structural reinforcement
Original Design	Steel	3.2–18.0	280.7	Stiffener on boom surface
New Design	CFRP	4.5–36.0	188.5	Rib inside of boom

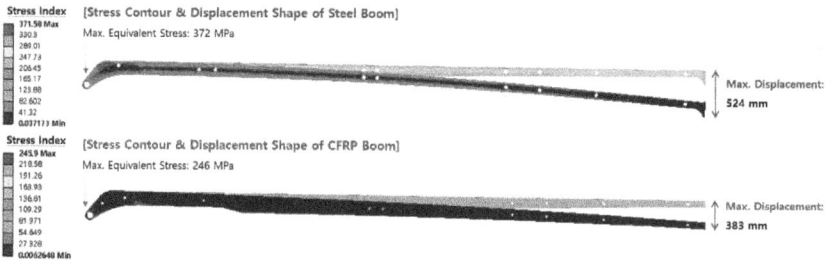

Fig. 9.9 Comparison of structural analysis result between steel end boom and CFRP end boom.[14]

Structural design, evaluation and component manufacturing of the CFRP boom have been performed. It has been confirmed that the CFRP end boom has a structural performance higher than that of the existing structure as shown in Fig. 9.9. It has also been confirmed that there was composite failure, through Puck's failure evaluation by FEA. Results of structural analysis of end booms are shown in Fig. 9.9. The result shows that CFRP end boom has better structural performance than the existing steel end boom under the same load conditions. Also, the evaluation results of fiber failure and interfiber failure by applying Puck's failure theory are shown in Fig. 9.10. The manufactured prototype components are shown in Fig. 9.11.

9.4.3 *Development of Elevator Cabin by Composites*

The development of lightweight elevators using aluminum composite material and CFRP was conducted. An aluminum elevator was designed by utilizing aluminum damped panel and aluminum sandwich structure, with 40% weight reduction in comparison with conventional steel elevators. Moreover, the CFRP elevator was designed by using CFRP on the elevator walls. Through this, it became conceivable to have a design that is 50% lighter than the existing steel elevator. The strength and stiffness of the designed elevator were verified by applying the load conditions given in the elevator inspection guideline. From the structural evaluation, it was confirmed that the newly designed elevator models had sufficient structural performance in the elevator inspection guideline. Design

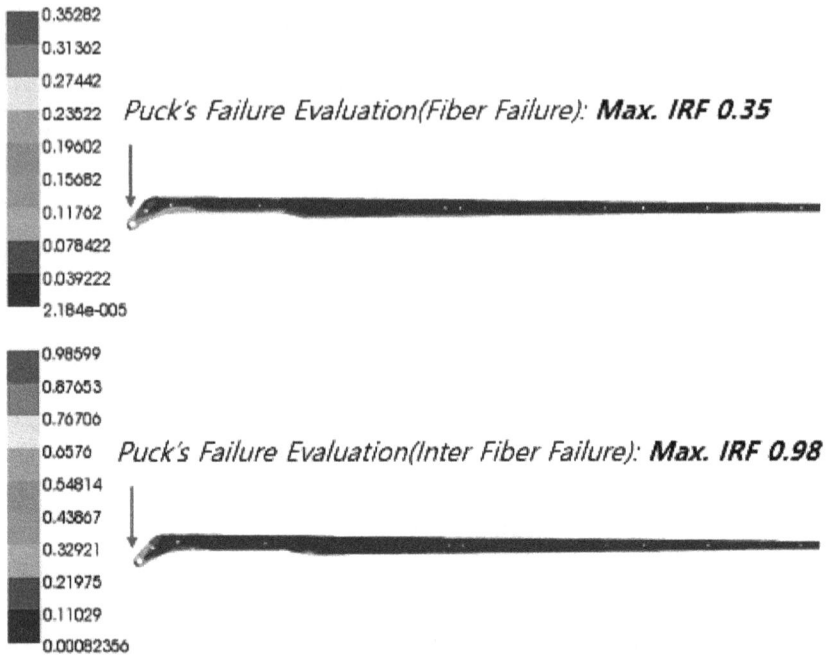

Fig. 9.10   Composite failure evaluation results of CFRP CPT end boom.[14]

specifications are summarized in Fig. 9.12, while the load and boundary conditions are shown in Fig. 9.13. Moreover, the structural analysis results are shown in Figs. 9.14–9.21, while the results of the design and evaluation are shown in Fig. 9.22.

9.4.4   Development of Elevator Rope by CFRP

CFRP was applied to reduce the weight of the elevator suspension belt. Conventional elevator suspension rope was heavyweight because it was made from steel. Elevator suspension belt applying CFRP was fabricated with 30% lighter weight and higher flexibility compared with conventional steel wire rope. A structural design of the belt applying CFRP was accomplished, and its structural integrity was evaluated by means of finite element method. Experimental verification was performed upon using manufactured CFRP

Fig. 9.11 Mold, manufacturing equipment and manufactured specimen.[14]

belts, with the FEA results showing the structure having sufficient structural integrity under design load conditions. It was confirmed that the performance target was satisfied through performance tests on the belt, and that the developed CFRP belt had sufficient competitiveness compared with the existing steel wire rope. In the future, design optimization could be accomplished, to reduce the

No.	Design Type	Weight (kg)
1	Steel Elevator	483.9
2	Al Elevator	293.1
3	CFRP Elevator	255.1

Ceiling
- Steel Elevator: Steel Structure
- Al Elevator: Al Sandwich Structure
- CFRP Elevator: Al Sandwich Structure

Car Wall
- Steel Elevator: Steel Sheet
- Al Elevator: DAL / Al 0.8mm+Al 0.8mm
- CFRP Elevator: CFRP / [45/-45/0/90/0/90],

Wall Reinforcement
- Steel Elevator: Steel Square Pipe
- Al Elevator: Steel Square Pipe
- CFRP Elevator: Steel Square Pipe

Car Frame
- Steel Elevator: Steel Structure
- Al Elevator: Steel Structure
- CFRP Elevator: Steel Structure

Platform
- Steel Elevator: Steel Structure
- Al Elevator: Al Sandwich Structure
- CFRP Elevator: Al Sandwich Structure

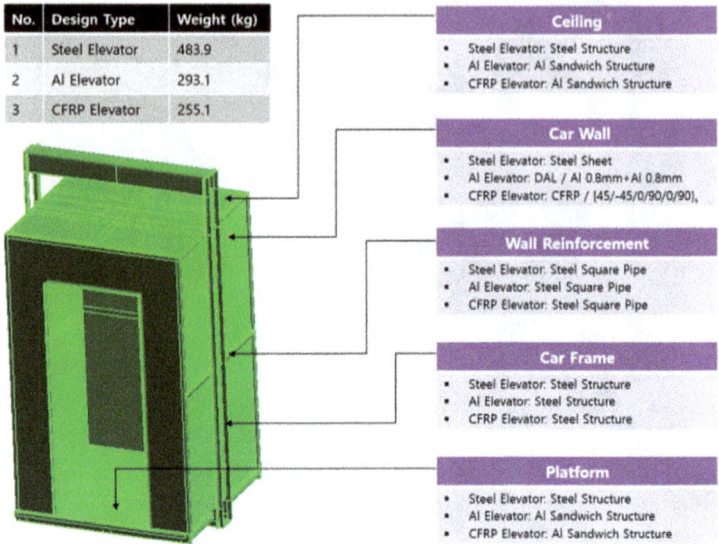

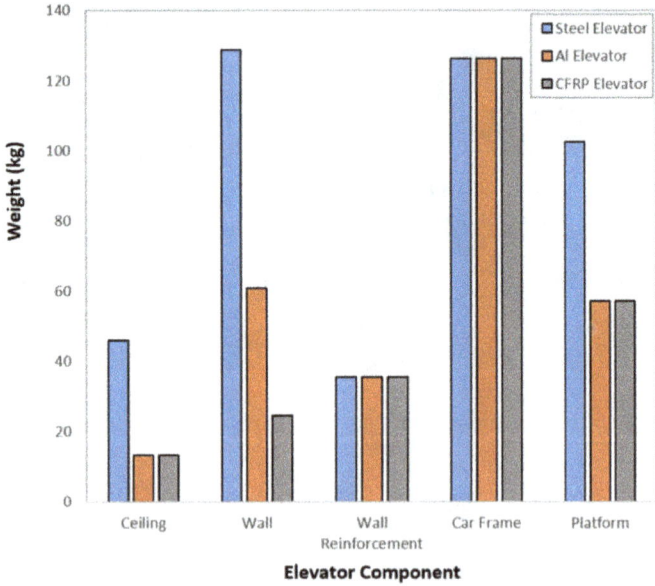

Fig. 9.12 Design specifications and weight comparison of elevator components.[15]

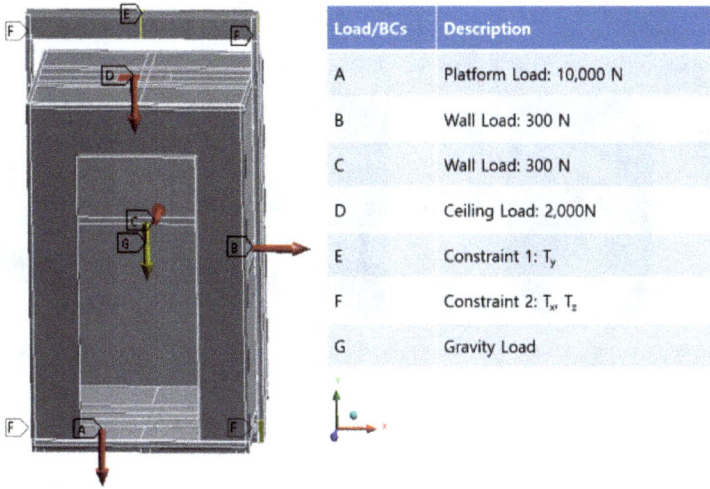

Fig. 9.13　Load and BCs of FEA of elevator models.[15]

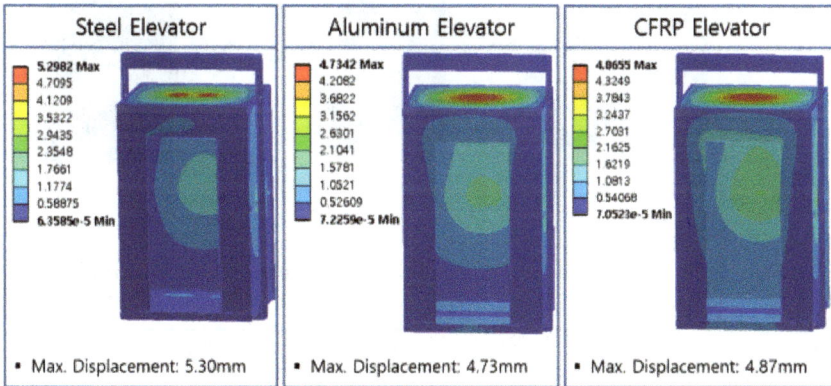

Fig. 9.14　Comparison of displacement of elevator full models.[15]

weight of the belt, and additional performance verification could be conducted. Comparison of design specification is shown in Fig. 9.23 and Table 9.4. Structural evaluation results by FEA and tests are shown in Figs. 9.24–9.27 and Table 9.5.

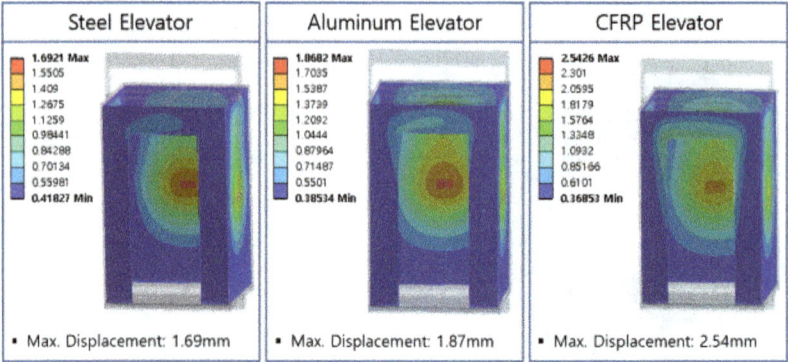

Fig. 9.15   Comparison of displacement of elevator wall components.[15]

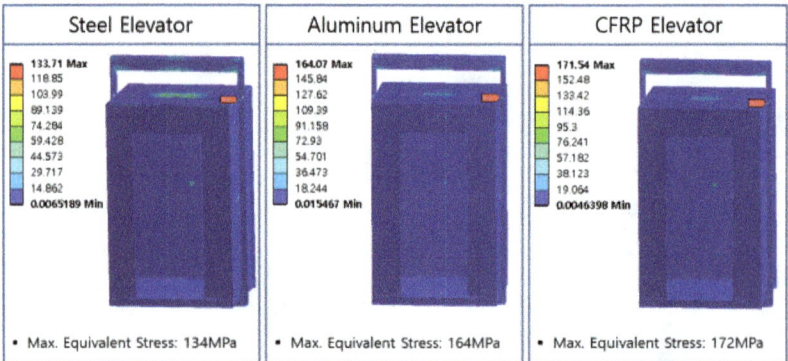

Fig. 9.16   Comparison of equivalent stress of elevator full models.[15]

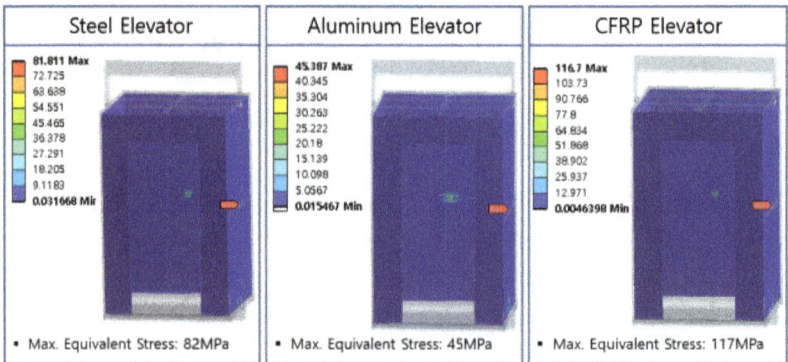

Fig. 9.17   Comparison of equivalent stress of elevator wall components.[15]

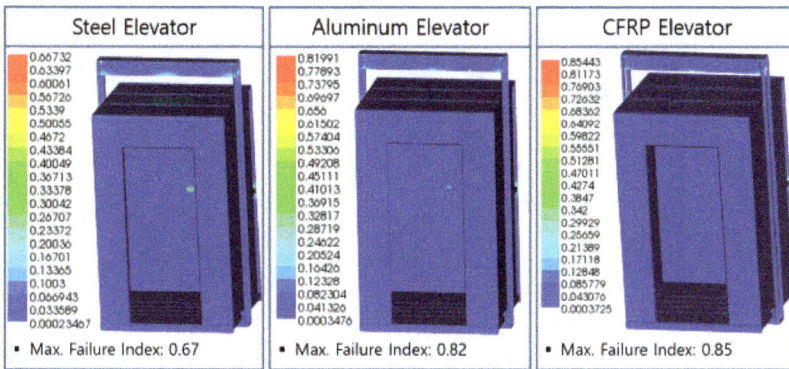

Fig. 9.18 Comparison of IRF of elevator full models.[15]

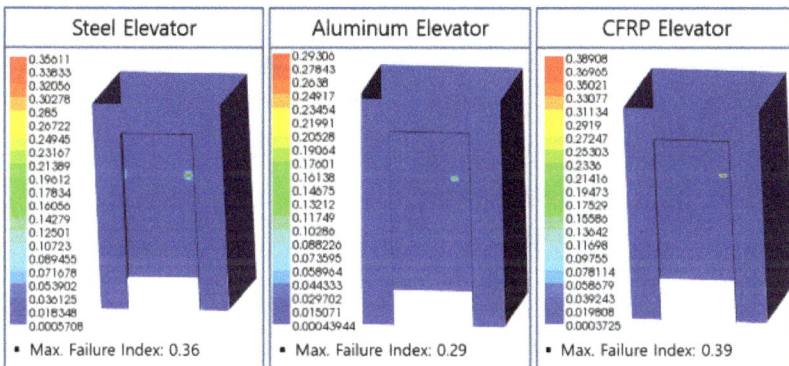

Fig. 9.19 Comparison of IRF of elevator wall components.[15]

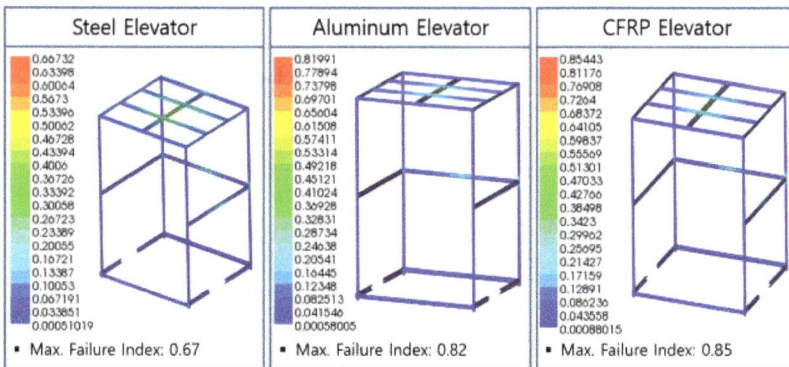

Fig. 9.20 Comparison of IRF of wall reinforcement components.[15]

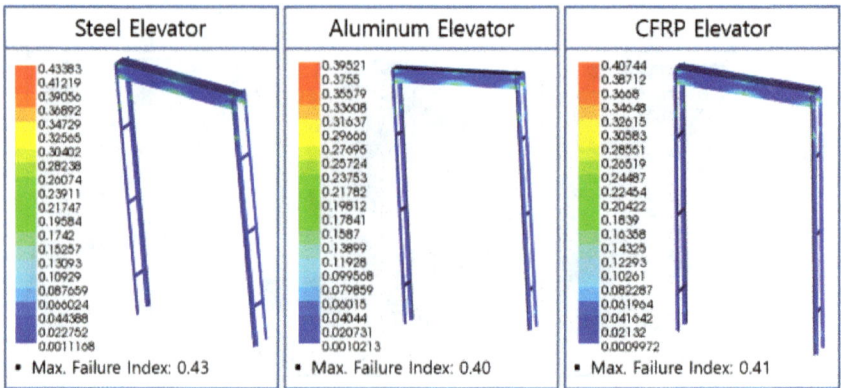

Fig. 9.21 Comparison of IRF of car frame components.[15]

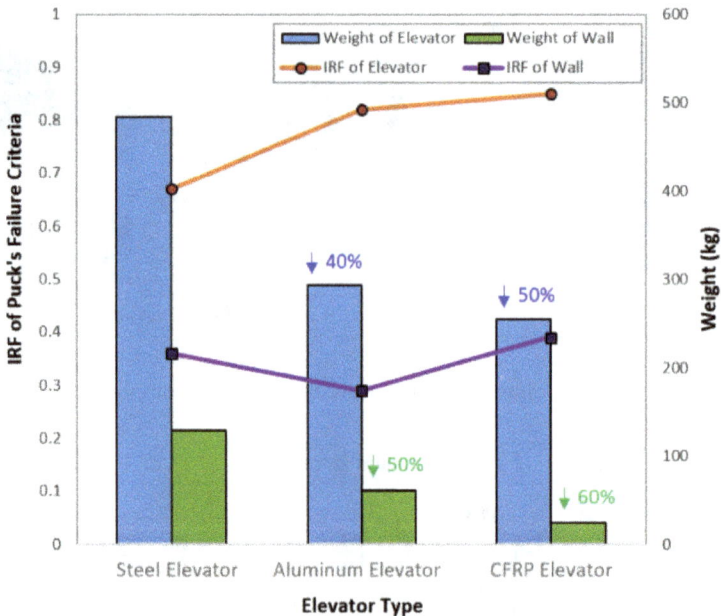

Fig. 9.22 Result of structural design and structural evaluations.[15]

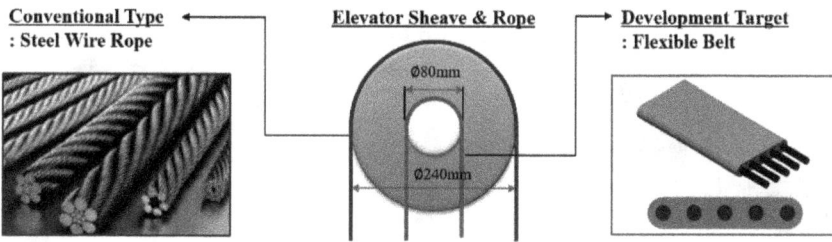

Fig. 9.23 Comparison of conventional elevator rope and flexible belt of high-rise elevator.[16]

Table 9.4 Comparison of design specification between conventional rope and developed belt for elevator[16]

Contents	Material	Weight (kg/m)	Thickness (mm)	Min. breaking load (N)	Passenger capacity	Number of ropes (EA)	Sheave diameter (mm)
Wire Rope	Steel	0.17	6	27,000	P 8–24	6–12	240
Developed Belt	CFRP	0.12	2	32,000	P 8–24	3–4	80

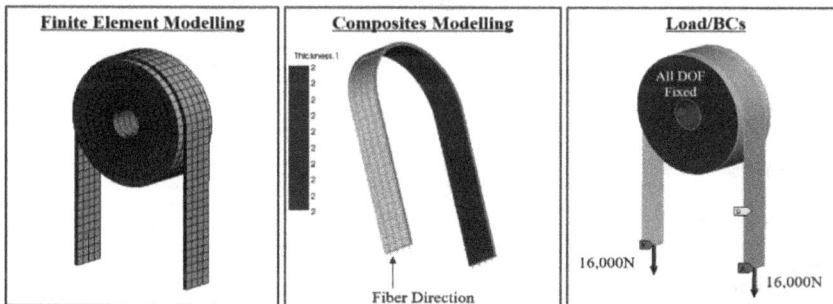

Fig. 9.24 Pre-process of FEA for CFRP suspension belt.[16]

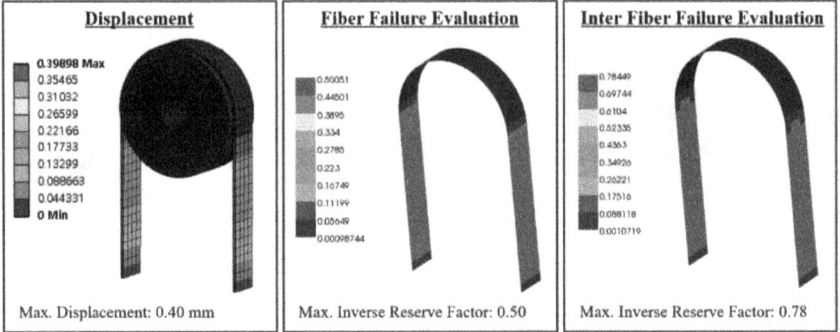

Fig. 9.25 Structural analysis results of CFRP suspension belt.[16]

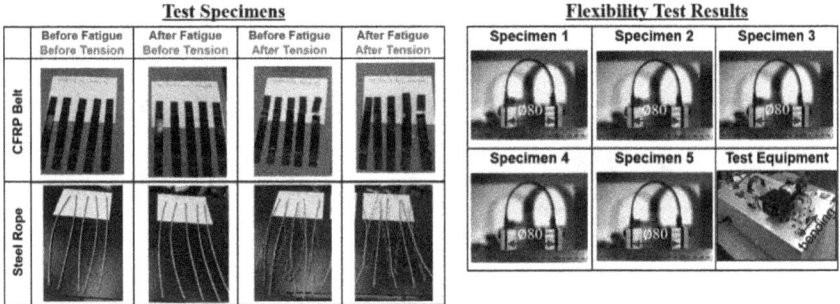

Fig. 9.26 Test specimens for tensile test and flexibility test and the flexibility test results of CFRP belt.[16]

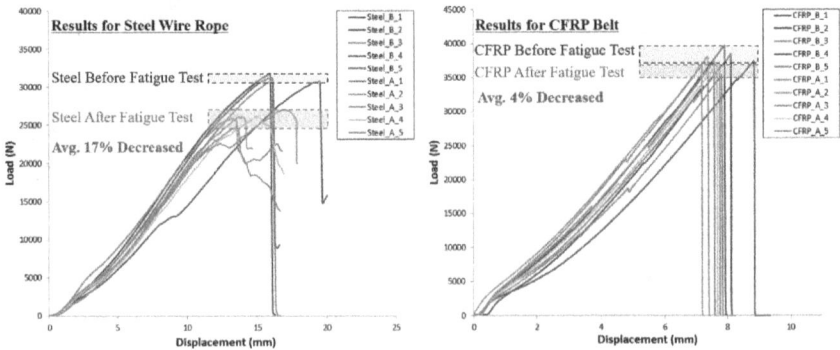

Fig. 9.27 Comparison of tensile test results between steel wire rope (left) and CFRP belt (right).[16]

Table 9.5 Performance test results of CFRP belt[16]

No.	Test items	Target	Results
1	Specific Weight	Less than 0.14 kg/m	O.K. (0.12 kg/m)
2	Tensile Strength	More than 32,000 N	O.K. (37,000 N)
3	Flexibility	Flexibility for Ø 80	O.K.
4	Friction Coefficient	More than 0.4	O.K. (0.6)
5	Thermal Resistant	More than UL94 V2	O.K. (UL94 V0)
6	Hardness	More than HS 85	O.K. (HS 86)
7	Durability	More than Steel Wire Rope	O.K.

9.5 Conclusions

In this chapter, applications of composite materials in automobiles and elevators have been presented. The theory and procedure of composite design in the automotive field are observed, and examples of parts development using aluminum and CFRP are described. The composite material is lighter than steel components and has excellent performance, making it possible to execute various things that were previously impossible. Currently, there are not many cases of composite parts' commercialization in the automotive and elevator industries due to their higher cost as compared to existing steel parts. However, their application should be gradually expanded in the field of automobiles and elevators through technological development. In particular, carbon fiber reinforced composite materials that can represent composite materials can replace existing steel and light metal structures because of features such as strength, superior heat resistance, durability and chemical resistance, which are more than 10 times stronger than the existing steel. Composite materials are already applied as core materials in areas where performance is a priority, such as in aerospace, aviation and defense. In certain industries including the automobile and shipbuilding/marine industries, composite parts are more expensive than the existing steel and metal parts. It is difficult to say that composite materials are widely used on various industries. It has been evaluated that overcoming the high price problem of the composite material parts through the development of materials/parts/equipment is feasible.

References

[1] A. K. Kheir, *Buildings*, 3, 2015, 1070.

[2] S. Suoranta, Tall Building and Elevators: A Review of Recent Technological Advances, Buildings 2015, 5(3), 1070–1104.

[3] A. N. Kolmogorov, Lightweight Ropes for Lifting Applications, in Proc. of the 2006 OIPEEC Conference, Athens, 1954, Vol. I, Greece, Athens, 2006, pp. 33–54.

[4] K. Y. Lee, B. S. Kong and H. P. Woo, *Trans. KSAE*, Vol. 6, No. 6, 166–173, 1998.

[5] B. Kim, W. Kim, S. Bae and Y. Lee, *Renew. Energ.*, 54, 2013, 166.

[6] S. Y. Bae and Y. H. Kim, *Mod. Phys. Lett. B.*, 33, 2019, 14–15.

[7] K. K. Jin, S. K. Ha, J. H. Kim and H. H. Han, *Korean Soc. Compos. Mater.*, 24, 2011, 10–16.

[8] B. Harris, *Engineering Composite Materials*, The Institute of Materials, 1999.

[9] F. C. Campbell, *Structural Composite Materials*, ASM International, 2010.

[10] A. Puck and H. Schurmann, *Compos. Sci. Technol.*, 58, 2002, 1045–1067.

[11] A. Puck, M. Kopp and M. Knop, *Compos. Sci. Technol.*, 62, 2002, 371–378.

[12] Ansys Inc., *Ansys Composite PrepPost User's Guide*, Ansys Inc., 2013.

[13] S. Y. Bae, H. C. Cho, H. G. Ahn and G. M. Bae, *Mod. Phys. Lett. B*, Vol. 34, No. 07n09, 2040035 (2020).

[14] S. J. Lee, I. S. Chung and S. Y. Bae, *Mod. Phys. Lett. B*, Vol. 33, No. 14n15, 1940033 (2019).

[15] S. Y. Bae, S. M. Yoon and Y. H. Kim, *Mod. Phys. Lett. B*, Vol. 34, No. 07n09, 2040032 (2020).

[16] S. Y. Bae and Y. H. Kim, *Mod. Phys. Lett. B*, Vol. 34, No. 07n09, 2040034 (2020).

www.ingramcontent.com/pod-product-compliance
Lightning Source LLC
Chambersburg PA
CBHW050539190326
41458CB00007B/1845